The Structure of Musical Sound

The Structure of Musical Sound

Willard Charles Sperry

Illustrator: Kurt Sperry

iUniverse, Inc.
New York   Bloomington

The Structure of Musical Sound

Copyright © 2009 Willard Charles Sperry

All rights reserved. No part of this book may be used or reproduced by any means, graphic, electronic, or mechanical, including photocopying, recording, taping or by any information storage retrieval system without the written permission of the publisher except in the case of brief quotations embodied in critical articles and reviews.

Slinky® is a registered trademark of Poof Slinky, Inc. Slinkys are available at toy stores and from scientific apparatus suppliers.

iUniverse books may be ordered through booksellers or by contacting:

iUniverse
1663 Liberty Drive
Bloomington, IN 47403
www.iuniverse.com
1-800-Authors (1-800-288-4677)

Because of the dynamic nature of the Internet, any Web addresses or links contained in this book may have changed since publication and may no longer be valid. The views expressed in this work are solely those of the author and do not necessarily reflect the views of the publisher, and the publisher hereby disclaims any responsibility for them.

ISBN: 978-1-4401-7507-7 (sc)
ISBN: 978-1-4401-7505-3 (dj)
ISBN: 978-1-4401-7506-0 (ebk)

Printed in the United States of America

iUniverse rev. date: 12/29/2009

Preface and Acknowledgements

Practically every university requires its students to take a science course. *The Structure of Musical Sound* is a text for such a course designed especially for arts and humanities students.

It is embarrassing to admit that it took me six years of teaching physics to realize that the normal first year physics courses were neither designed to be general education courses nor did they satisfy students who wanted a general-ed science course that did not take for granted that its students were fluent in the language of mathematics. For these students our first year courses were difficult bilingual exercises: an often unfamiliar, technical English combined with mathematics.

So, I began to design the course, The Physics of Musical Sound, which would overcome these objections. There were not many texts available; and most used a math-logic style that was fine for me who knew the physics and wanted the musical sound information, but were not what I wanted for my students. Luckily I found *Horns, Strings, and Harmony* by Arthur H. Benade and quickly adopted it; and it just as quickly went out of print. I had no choice but to write my own text, at first as lecture notes that I gave to my students. Then, as experience and trial-and-error firmed up my ideas, I decided to prepare a book. *The Structure of Musical Sound* is the result.

All the other musical sound texts that I know of have also been written by physicists, and taught in Physics Departments. This gives

Preface and Acknowledgements

them the necessary scientific rigor and attention to principal ideas; but, as I said, often uses a mathematical elegance that is often neither understood nor appreciated by the reader whom I have in mind. The first time I taught musical sound I cavalierly wrote an equation on the blackboard and started to expound on its features. Two students got up, walked out and never came back. *The Structure of Musical Sound* is written to avoid this problem. Of course, it uses mathematics, mostly arithmetic and high school algebra, and even this is introduced in a gradually increasingly demanding way so that the reader can begin at the shallow end of the pool. Every time he or she meets a new type of mathematical or logical technique, it is accompanied by a thorough explanation and a detailed example of its use. QUESTIONs are placed throughout the text, closely following the material about which they ask. There are not too many of them, and the student is expected to answer them all. They are designed to illuminate the text's ideas, and to give the reader a mini test about whether he or she understands the material well enough to be able to use it. The QUESTIONs are graded also, at first requiring only simple arithmetic calculations and drawing conclusions from the resulting numbers, advancing to the use of algebra to combine formulas, and then discussing the meaning of the resulting combination. The answers with their accompanying solutions are presented at the end of the book.

This book could have an additional, more general readership. Musicians could read it to learn more about their instruments, and computer programmers or engineers could use its descriptions of acoustic problems and their solutions as development ideas for computer-controlled audio systems. I have made *The Structure of Musical Sound* both instructional and interesting, and to be read with pleasure by educated persons with general interests in science and/or music.

Merryl Drakard read an early draft of *The Structure of Musical Sound* and contributed many valuable comments and suggestions for improvements. I thank her very much for this assistance.

Preface and Acknowledgements

JoAnne Sperry provided technical information and hands-on assistance during the preparation of the text.

Vivian Goffart told me a lot about playing the flute, and how to make audio and video recordings.

Virginia Benade Belveal kindly helped me acquire permission to use an illustration from *Horns, Strings & Harmony*.

Additional Preface for Teachers

A university course using this text could be done in a semester or one or two quarters. It should include the actual DEMONSTRATIONs, either in the classroom or in a weekly laboratory session where the students can do these or other investigations the instructor will undoubtedly know. The laboratory work is important: it requires the students to assemble and manipulate apparatus for measuring sound, it requires them to use this apparatus to measure sound, and it requires them to use the measurements to answer questions and explain phenomena in a lab report written after each lab session. All of these are necessary parts of doing science.

A set of my lab instructions to the students is available on the webpage http://thestructureofmusicalsound.com. With a few exceptions they require the usual apparatus of freshman physics laboratory work. These labs take two hours or less, and another hour or so to write the report.

The mathematical level of this text and its course, which is rather simple, can be increased by the instructor as applicable to the particular course and students. Certainly the instructor can add fine mathematical expositions to the down-to-earth physical models I've provided. However, not all the students for whom this book is written will enjoy such elegance.

Table of Contents

Preface and Acknowledgements v
Additional Preface for Teachers. ix
List of Illustrations. xv
List of DEMONSTRATIONS xvii
List of Recordings xix
Introduction xxi

PART 1: RULES

CHAPTER 1. PROPERTIES OF SOUND 3
  1.1 Sound in General and a Model 3

CHAPTER 2. MUSICAL SOUND 17
  2.1 What is Musical Sound?. 17
  2.2 Pitch and Frequency 31

CHAPTER 3. MUSICAL VIBRATIONS AND THEIR
                  VIBRATORS 39
  3.1 Vibrating Strings. 39
  3.2 More Vibrating Strings, Preventing Modes, Combining Modes . 48
  3.3 Complex Vibrations in a String 49
  3.4 Combinations of Musical Sound Waves. 51
  3.5 Beats 59
  3.6 Ears Create Additional Frequencies 62
  3.7 Analysis of Two Sounds 63
  3.8 Pitch-Frequency, Loudness-Sound Pressure, Timbre-Mode
      Content 71

CHAPTER 4. MUSICAL SCALES AND TEMPERING 75
  4.1 Piano Design: Good-Bye, Just Tempering; Hello, Even
      Tempering. 75

Contents

PART 2: MUSICAL INSTRUMENTS

CHAPTER 5. STRINGED INSTRUMENTS--MAKING THE SOUND 83
5.1 Loudness, The Need for Soundboards and Soundboxes 83
    5.1.1 Soundboards and Soundboxes. 84
5.2 Piano 86
5.3 Viols and Violins. 90
5.4 Guitar and Banjo. 93

CHAPTER 6. SOME MUSICAL PERCUSSION INSTRUMENTS . . 99
6.1 Membranes and Thick Strings 99
6.2 Music Directly from Mathematics 103

CHAPTER 7. WIND INSTRUMENTS--MAKING THE SOUND . . 105
7.1 Standing Waves in Air 105
7.2 Boundary Conditions and the Modes of Ideal Cylindrical Cavities. 114
7.3 Sustaining the Tone. 123
7.4 Pressure Pulses in Long Cavities and How They Determine the Modes. 127
    7.4.1 Cylindrical Cavity with a Reed at One End and the Other End Closed. 128
    7.4.2 Cylindrical Cavity with a Reed at One End and the Other End Open 130
    7.4.3 Conical Cavity and Cylindrical Cavity Open at Both Ends 136
7.5 Real Reeds 139
    7.5.1 Real Puffs 142

CHAPTER 8. BRASS INSTRUMENTS. 145
8.1 Ideal Conical Brass Instruments. 145
8.2 Real Brass Instruments 148
    8.2.1 The Bell 151

CHAPTER 9. WOODWINDS. 155
9.1 Ideal and Real Woodwinds 155

Contents

CHAPTER 10. TWO OTHER KINDS OF WIND INSTRUMENTS 159
10.1 Harmonicas, Accordions, Concertinas 159
10.2 Ocarina 161
    10.2.1 Helmholtz Resonators in Other Woodwinds 165

PART 3: ACOUSTICS

CHAPTER 11. ROOM ACOUSTICS 169
11.1 The Language of Musical Acoustics 171
    11.1.1 Musical Descriptions 171
    11.1.2 Scientific Descriptions 173
        11.1.2.1 Reflection and Resonance 174
        11.1.2.2 Reverberation Time 188
        11.1.2.3 Delay Time 192
11.2 Active Acoustics 194

APPENDICES

APPENDIX A. SOUND BECOMES LESS LOUD: A CLOSER LOOK AT THE STRUCTURE OF A GAS 211
A.1 A Model of a Gas 211
A.2 The Ideal Gas Law 215
A.3 The Kinetic Theory of Gases 217
A.4 Sound Changes Loudness 226
    A.4.1 Sound in Air Alone 226
    A.4.2 Sound Hits Obstacles--Reflection, Diffraction, Absorption 227
        A.4.2.1 Reflection 227
        A.4.2.2 Diffraction 228
        A.4.2.3 Absorption 229
        A.4.2.4 Impedance 233

Contents

APPENDIX B. COMPUTER-CONTROLLED AUDIO ELECTRONICS 235
  B.1 The Computer 238
    B.1.1 Analyzing the Sound. 239
      B.1.1.1 The Fourier Transform 239
      B1.1.2 Sampling the Sound 240
    B.1.2 Modifying the Sound 249
      B.1.2.1 Changing Loudness and Reverberation Times, Reflected Sound. 249
      B.1.2.2 Delay Times. 256
  B.2 Disclaimer 257

APPENDIX C. SCIENCE SYMBOLS: PHYSICAL QUANTITIES, PHYSICAL OBJECTS, AND HOW THEY ARE WRITTEN 259

Answers To QUESTIONS 263
Notes 313
Glossary of Technical Terms 317
Additional and Extended Readings 331
Sources of Illustrations 333
Index 335

Illustrations

Figures

Figure 2.1 Ripple Tank Waves.............................. 27

Figure 2.2 Pitch Ranges of Voices and Musical Instruments......... 29

Figure 2.3 Piano Keys 32

Figure 3.1 Sound Pressure Amplitude vs. Frequency 72

Figure 5.1 The Piano 88

Figure 7.1 Some Wind Instruments According to Their Reeds 140

Figure 8.1 Range of an Ideal Three-valve Conical Brass Instrument Whose Pedal Note, f, is C_3........................ 147

Figure 8.2 A Type of Tenor Trombone......................... 154

Figure 11.1 The Original Teatro alla Scala 170

Figure 11.2 St. Martin's Hall................................ 170

Figure 11.3 Standing Waves Between Walls a Distance L Apart..... 174

Figure 11.4 Listener and Standing Waves 178

Figure 11.5 Standing Waves Divided into $\lambda/16$ Wide Regions...... 179

Figure 11.6 Sound Paths in an Auditorium 192

Figure 11.7 Delay Time and the Perception of Sound............. 193

Figure 11.8 Block Diagram of Normal and Computer-changed Sound From Source to Listener.................... 197

Figure A.1 Diffraction of Water Waves 228

Figure B.1 Block Diagram of Normal and Computer-changed Sound From Source to Listener.................... 237

Illustrations

Figure B.2 Block Diagram of Computer Making Corrections to the Sound. 246

Figure B.3 Bandpass Filters Separate the Frequencies 247

Figure B.4 Initial Loudness and Reverberation Times 251

Figure B.4.A Added Loudness to Cause Longer Reverberation Time. 252

Figure B.5 Block Diagram of Computer Correcting Reverberation Times. 253

Tables

Table 3.1 Mode and Wavelength for a Length L Vibrating String. 44

Table 3.2 Modes and their Musical Intervals. 47

Table 4.1 The Variations of the Pitch a_4. 80

Table 11.1 Resonant Modes in a Room and in an Auditorium 176

Table 11.2 Wavelengths and Frequencies that Could Cause Bad Seats. 184

Table A.1 Absorption Coefficients of Various Surfaces and Objects. . 232

Table B.1 Sampling Times and ADC Values About One Second After Music Starts 241

Table B.2 Sampling Times and Frequency Amplitudes. 248

Chart

Chart 2.1, Semitone Intervals in Major and Minor mode Scales. 34

DEMONSTRATIONS

DEMONSTRATION I: Sound, Vibrations 4

DEMONSTRATION II: Sound Made Visible 5

DEMONSTRATION III: Operation of a Simple Microphone 7

DEMONSTRATION IV: Making the Tuning Fork's Vibrations Visible . 10

DEMONSTRATION V: Visible Traveling Waves 11

DEMONSTRATION VI: Elasticity of Air 13

DEMONSTRATION VII: Air is Necessary for Sound 14

DEMONSTRATION VIII: Musical Sound and Noise 17

DEMONSTRATION IX: Loudness and Pitch Displayed on an
    Oscilloscope 21

DEMONSTRATION X: Traveling Waves Intersect, Superposition . . . 25

DEMONSTRATION XI: Standing Waves 39

DEMONSTRATION XII: Ghostly Combination Frequencies 70

DEMONSTRATION XIII: Soundboard 83

DEMONSTRATION XIV: Modes, Nodes, and Antinodes, Chladni
    Figures 84

DEMONSTRATION XV: Tapped Sound Changes When Block is
    Supported at its Node 102

DEMONSTRATION XVI: Standing Longitudinal Waves 106

DEMONSTRATION XVII: Modes of a Tube Open at Both Ends . . . 121

DEMONSTRATION XVIII: Microphone Probes Tube to Locate
    Nodes and Antinodes of its Resonances 122

DEMONSTRATIONS

DEMONSTRATION XIX: Sustaining an Oscillation 125

DEMONSTRATION XX: Construct a Cylinder with a Reed, and Play It 139

DEMONSTRATION XXI: Open Side Holes Change Length of Cavity 157

DEMONSTRATION XXII: Resonance in a Bottle 162

DEMONSTRATION XXIII: A Blown Helmholtz Resonator 163

DEMONSTRATION XXIV: Mechanical Model of a Gas 213

DEMONSTRATION XXV: Hard and Soft Collisions 230

Recordings

These recordings can be heard on the web page
www.thestructureofmusicalsound.com

Band 1.   Section 1.1, DEMONSTRATION VII, (Air is Necessary for Sound)

Band 2.   Section 2.1, DEMONSTRATION VIII, (Musical Sound and Noise)

Band 3.   Section 2.1, DEMONSTRATION IX, (Loudness and Pitch Displayed on an Oscilloscope), Single Pitch Changing Loudness

Band 4.   Section 2.1, DEMONSTRATION IX, (Loudness and Pitch Displayed on an Oscilloscope), Tones Changing Frequency

Band 5.   Section 2.2, Piano Notes D_4 E_4 F_4 G_4 A_4 B_4 C_5 D_5

Band 6.   Section 2.2, Intervals: MAJOR THIRD, PERFECT FOURTH, PERFECT FIFTH

Band 7.   Section 3.4, Complex Tones: Timbre Differences in Two Intervals

Band 8.   Section 3.4, PERFECT FIFTH: 200Hz and 300Hz

Recordings

Band 9. Section 3.4, Complex Tone: 300Hz and 400Hz

Band 10. Section 3.5, Beats

Band 11. Section 3.5, Beats Become Tones

Band 12. Section 3.5, Beats From 400Hz and 402Hz

Band 13. Section 3.5, Beats From 300Hz and 305Hz

Band 14. Section 3.7, Two Complex Tones: 300Hz and 400Hz; 200Hz and 300Hz

Band 15. Section 3.7, Complex Tones: 500Hz and 612Hz

Band 16. Section 3.7, DEMONSTRATION XII, (Ghostly Combination Frequencies)

Band 17. Section 3.8, Successive 100Hz Increases

Band 18. Section 4.1, Just and Even Tempered Tones

Band 19. Section 5.1, DEMONSTRATION XIII, (Soundboard)

Band 20. Section 6.1, DEMONSTRATION XV, (Tapped Sound Changes When Block is Supported at its Node)

Band 21. Section 7.2, DEMONSTRATION XVII, (Modes of a Tube Open at Both Ends)

Band 22. Section 9.1, DEMONSTRATION XXI, (Open Side Holes Change Length of Cavity)

Band 23. Section 4.1, Just and Even Tempered Intervals

Introduction

The explanations of musical sound, of the instruments that produce it and of room acoustics presented in *The Structure of Musical Sound* (TSMS) are elementary but not always simple. These explanations are, or are the direct results of, the physical laws you might also learn in the lectures, demonstrations, and laboratory sessions of a first year university physics course. But, TSMS assumes that you know only arithmetic and a bit of algebra. TSMS does require you to learn and to use the analytic and problem solving skills taught in university level science courses.

Not all of the many Figures, Graphs, and Tables in TSMS are numbered. They should be read when encountered, just as you would read and understand text. The numbered ones are referred to in other places in this book, and numbers make them easier to find.

QUESTIONS are inserted in the text. They should be answered as they come. This will help you determine whether or not you have understood the previous material well enough to be able to use it. The solutions to these QUESTIONS are presented in the "Answers to the QUESTIONS" section. With these you can test yourself and/or also see examples of analytic and problem solving skills.

The Structure of Musical Sound uses some science terminology and science symbols. The "Glossary of Technical Terms" provides scientific meanings of words that also have musical or more everyday uses.

Introduction

However, even the scientific meanings there have been restricted to be pertinent to musical sound.

Appendix C. "Science Symbols" presents and explains the symbols used for Physical Quantities and Physical Objects. Metric system units are used wherever appropriate. If you are not familiar with them this book will help you learn them.

You, the reader of TSMS, can become the listener of some of the musical sounds described in it by visiting a web site. The "List of Recordings" has the details.

PART 1: RULES

CHAPTER 1.
PROPERTIES OF SOUND

1.1 Sound in General and a Model

We begin by studying sound in general, identifying and explaining its properties. The particular kind called musical sound has all the general properties; but, as you will see, some of them will be limited to specific values or will lie within a narrower range of the complete scope of sound.

Little experiments, or demonstrations, will show you some of the general properties. These demonstrations use apparatus that extends our ability to perceive the structure of sound beyond the information that our ears alone can provide. The apparatus will let you observe the very quick changes of very small things that are basic to sound, and show you the characteristics of air which make sound possible. Sound can pass through other media. You can hear under water, but musical sound travels mostly through air, and this will be the playing field on which our exploratory game takes place.

The demonstrations have a twofold purpose: one, to present and explain the physical properties of sound; and two, to let you develop a correspondence between what you hear and these properties. The demonstrations and the explanations that accompany them will often be suggestive rather than conclusive; but they will make sense. After several demonstrations a model of sound will be proposed. The model will be a description of a thing that has these properties, and will be our description of the physical structure of sound. Model making is one of the most speculative, important, and pleasant parts of science. Once a model of sound is suggested we will realize that if the model is correct sound must have additional properties or must behave in certain ways. More demonstrations and experiments can be

Part 1: Rules

mounted to check these predictions. In this way you can gain confidence, or not, that the model is correct.

I admit that the model proposed here will be the accepted model of sound, and that the demonstrations have been chosen to show its properties. Other models of sound, which looked promising, or even good but which were discarded because they finally did not agree with experiment, are not mentioned. This can tend to make science look like an arrow that always hits the bull's-eye, which is certainly not true.

The model proposed and supported by the demonstrations is that sound is a train of high and low pressure regions moving through the air.

The demonstrations will also show that vibrating objects cause these high and low pressure regions. And they will show that musical sound has unique properties. I have arranged the demonstrations in order that one suggests the next, and so they tell a story.

DEMONSTRATION I: Sound, Vibrations

Apparatus: Tuning Fork

Strike a tuning fork on a soft surface such as your knee. You will hear its soft tone. With your other hand gently touch the tines and feel the buzzy vibration. The sound ceases when the vibrations do. The conclusion is that the vibrating tines cause the tone.

It's impossible to tell in detail what's happening, though. The vibrations are too quick and too small to follow with sight or touch. Sound is also invisible. You did hear something, however, and the next demonstration adds some instrumentation, which will make some of the features of sound visible.

DEMONSTRATION II: Sound Made Visible

Apparatus: Tuning Fork, Microphone, Oscilloscope

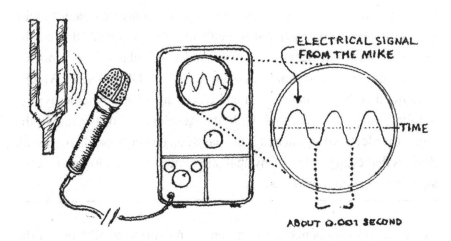

Strike the tuning fork. Along with the sound, a pattern appears on the oscilloscope's screen. The microphone (mike) transforms the sound into a changing electric signal, which the oscilloscope (scope) displays as a graph of the signal's strength vs. time.

The oscilloscope begins the graph when the signal has a specified height and slope. In this way, if the signal has a reoccurring form the scope will continue to redraw the same graph in the same place on the scope's screen, and you will see it as a fixed pattern. Of course, the signal changes with time, but the scope waits until the signal has the right height and slope to begin graphing. If the sound does not have a repeating form the scope will begin the graph almost at random, and the display will be continually changing and impossible to read or to understand.

So, if the sound has a repeating form, the scope will be able to show it in a steady unchanging display. This is happening for the sound from the tuning fork; the display repeats a basic shape, called a cycle, about every one thousandth of a second. The electric signal from the mike is weak, and the scope has amplifiers to make the display big enough to read and measure. You can adjust the display's time scale and amplification; these capabilities are built into the oscilloscope and make it one of the most useful scientific instruments for showing the properties of a large range of electric signals. Many companies make devices to change physical phenomena into electric signals that can be displayed on oscilloscopes or stored and manipulated in computers. Microphones are good examples of these devices. DEMONSTRATION II shows that they change sound into electrical signals. Another type is the fuel gauge in an automobile. It changes the level of the fuel in the tank to an electrical signal that is displayed on the dashboard. In general, devices that change purely physical phenomena into analogous electric signals are called transducers.

There might be a connection between the repetitive oscillations of the fork's tines and the cycles on the scope's display. If this is so, and it is, what is the link between them? It is the sound generated by the fork and received by the mike. This should fix your attention on what's in the space between them, but you still don't know what is happening there, or what is traveling from the fork to the mike, other than that its name is sound.

What is the microphone changing into an electric signal? A demonstration taking a closer look at the microphone will answer this.

Chapter 1

DEMONSTRATION III: Operation of a Simple Microphone Apparatus: Microphone

This microphone is not the best; it was chosen because it is simple and is easily taken apart. The diaphragm is a thin metal disc held in place by the attraction from the two poles of the magnet. Metals conduct electricity.

A conductor moving near a magnet will have an electric signal generated in it whose strength and polarity (the plus or minus of the signal) depend on whether the conductor is moving fast or slow and whether it is getting closer or farther away from the magnet. This is the signal the scope displays and if you remove the diaphragm from the mike there will be no display on the scope. You should gently push on the diaphragm when it is back in place and see the display change. From this evidence you can correctly conclude that the sound is moving the diaphragm, and since the electric signal has alternating polarity (it goes above and below the zero level) the sound both pushes and pulls on the diaphragm.

The demonstrations so far tell you what sound does but not what it is. Can we invent a model that does both? Yes. Try this explanation.

    A. The vibrating tines create high and low pressure regions in the nearby air. By "high and low" it is meant just slightly more and less than normal atmospheric pressure. The low pressure is a partial

Part 1: Rules

vacuum. Appendix A describes why packing more air molecules into the same volume (such as adding air to a tire) raises the air pressure inside the tire.

B. These high and low pressure regions flow outward with the speed of sound from their original locations.

C. When a portion of these flowing regions intercepts the diaphragm they push (the highs) or pull (the lows) on it. This causes the vibration of the diaphragm.

Let's look at each of these steps in detail.

A.

THE TINES SQUEEZE THE AIR (PACK THE AIR MOLECULES INTO A SMALLER VOLUME) THIS CREATES A HIGH PRESSURE REGION

NOW THE TINES CREATE A BIGGER VOLUME FOR THE AIR MOLECULES. THIS CREATES LOW PRESSURE

Note that this creation of successive highs and lows is taking place at the same rate that the tines are vibrating: quickly. Also note, for future use, that any vibrating object will produce highs and lows next to it. Loud speakers are probably the most common example, but they are not musical instruments. Appendix A. "Sound Becomes Less Loud: A Closer Look at the Structure of a Gas" tells how you can also change the pressure by changing the air's temperature, but this method is not used to create musical sound.

Chapter 1

Although tuning forks are almost perfect examples of sound producing vibrators, there are lots of others; and Chapters 5 through 11 will describe musical ones. You will see that they all vibrate to produce highs and lows, and that they come in as many shapes as there are musical instruments or loud speakers.

B. An original high-pressure region will expand (excess air molecules will flow outward). This reduces the original high to normal atmospheric pressure, but creates a "halo" of high pressure around the original region.

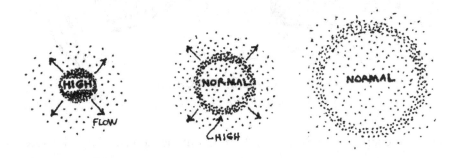

Here is a "skin" of high pressure expanding from the original high. The fixed number of excess air molecules in the original high fills an increasing volume skin and the high pressure becomes less as the skin grows. This also explains why sound gets less loud as it moves away from its source. With an original low-pressure region just reverse the direction of the molecules' flow. They flow into the low, leaving an expanding skin of low pressure.

The result is successive skins of high and low pressure expanding from the vibrating object. These are the sound waves. How fast do they move? With the speed of sound! Your ear intercepts a part of these skins and your eardrum vibrates like the mike's diaphragm.

Do the fork's tines really vibrate this way? Do they come together and then fly apart as shown or do they both move left and then both move right in which case the highs and lows would not be formed so efficiently? A demonstration, which illuminates the tines with a high-speed strobe light, will answer this.

DEMONSTRATION IV: Making the Tuning Fork's Vibrations Visible

Apparatus: Tuning Fork, Strobe Light

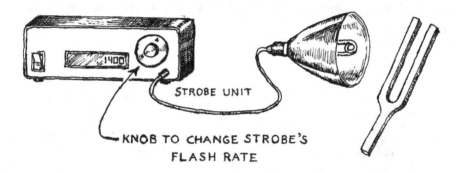

Adjust the strobe's flash rate until it is about the same as the fork's rate of vibration and you will see a "slow motion" picture of the fork. It is slow enough that you can easily see the tines vibrating as described in step A.

If you understand how this slow motion effect occurs you will be well on your way to understanding the sonic phenomenon called beats. In fact the slow motion effect you are seeing are visual beats. Here's the explanation.

If the strobe's flash rate were equal to the fork's rate of vibration you would see the tines at rest because they would be in the same place each time that the strobe flashed. If these two rates are not quite equal, however, the strobe will light the tines at progressively earlier (or later) times in its cycle of vibration. This will show the tines in slightly earlier (or later) position than shown with the previous flash. As the flashes continue you see the tines slowly move through the positions of their motion.

Chapter 1

This explanation of a sound wave propagating by means of the flow outward (or inward) of the excess air molecules is correct, but can easily lead to a misconception which must be corrected. The misconception is that the original high pressure's excess molecules continue to flow outward. In fact they go a very short distance before they collide with normal concentration air molecules. The original high pressure molecules collide with and bat the normal ones into having an excess concentration of high pressure. These in turn strike the next outer ones and the high pressure expands outward in a kind of 3-dimensional domino theory. This is impossible to see, and your ears do not sense that this is happening. However, this type of process occurs in other media. They are not air, but you can see them.

DEMONSTRATION V: Visible Traveling Waves

Apparatus: Rubber Tubing, Slinky

A rubber tube is stretched and then released. The stretch travels down the tube.

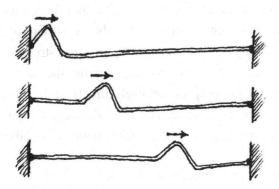

When the stretched tube is released the deformation travels across through the tube. The motion of any part of the tube is up and down as the deformation arrives and passes. None of the tube moves to the right.

Part 1: Rules

Again, here is another kind of medium with deformations that are not perpendicular to the direction of their travel. A Slinky is compressed at one end and released. The compression travels down the Slinky.

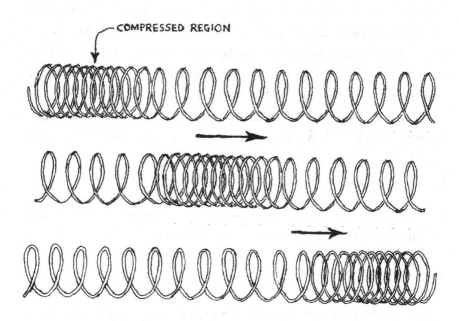

In both of these examples, the stretched or compressed parts correspond to a low or a high pressure region in air. Neither the stretched piece of tubing nor the originally compressed coils travel horizontally. If you tied a piece of red yarn to the top of the original stretch, or to one of the compressed coils, it would not move across the tube or across the Slinky. The stretch or compression itself travels, and the tube or Slinky only passes on this disturbance. The air molecules in the highs or lows also don't move much at all. If they did you would feel the wind.

Did you notice in this demonstration that in order for the disturbance to travel the medium must be elastic? Is air elastic? This must be so, if the tube and Slinky results can be applied to air. Also, the elasticity of air was assumed while explaining how the vibrating tines could create high and low pressure regions; i.e., the air molecules were contained in a different volume and then sprang out of (or into) it. The required elasticity of air can be demonstrated.

Chapter 1

DEMONSTRATION VI: Elasticity of Air

Apparatus: Large Syringe

SYRINGE WITH PLUGGED END

Push down on the plunger and feel the pressure build up as the volume of air decreases. The number of air molecules didn't change (no leaks). What did change was the number of molecules per volume, i.e. the pressure inside the syringe. When the plunger is released, the increased pressure pushes it upward and the air again attains its original volume. This air is acting like a spring; it has elasticity.

C.

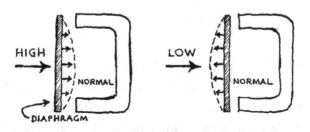

Part 1: Rules

High and low pressure skins (the sound's wave fronts) hit the diaphragms as shown above. The high pressure pushes the diaphragm to the right, and the low pressure lets the higher, normal atmospheric pressure push it to the left. There is no pull, just pushes from the opposite direction.

Therefore, the model of sound, being that of moving high and low-pressure regions in the otherwise still air, seems to agree with all of the demonstrations. Of course this doesn't make it right. There might be hundreds of other models of sound, which could also agree. Is there any way to test whether our air-pressure based model is wrong? Yes! According to our model, sound cannot exist where there is no air. Bring on the vacuum pumps!!

DEMONSTRATION VII: Air is Necessary for Sound

Apparatus: Vacuum Pump, Sound Source in a Bell Jar

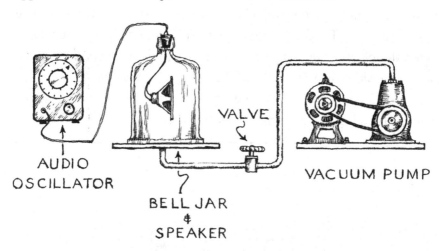

The audio oscillator produces a cycling electric signal, which the loud speaker plays as a tone. It has dials with which you can change the pitch and loudness of the tone. As its name implies, the range of pitches it can generate is about equal to the range you can hear, i.e. audio. The vacuum pump pumps the air out of the bell jar. The valve is a little more

Chapter 1

complicated than shown. It can not only connect or isolate the pump from the bell jar, but it can let air back into the pumped bell jar.

Start with air in the bell jar and with the loud speaker playing a tone you can hear. Then, begin to pump the air out of the jar. The sound gets fainter (less loudness but no pitch change) as the air is pumped. When the air is let back in, the sound returns. Listen to the recording of this happening (Band 1)[1].

So, air is necessary for sound. This alone does not prove our model correct, but it does add new evidence in its favor.

You should now have some confidence that the claim that sound is created by vibrating objects is correct, and that the model I have proposed is satisfactory.

QUESTIONS, such as the one below and the others that you will encounter, are an important part of this book. Don't just continue to read on when you come to them, but take time to answer them. Some will not have a unique answer. You will be required to state some answer and then defend it. A scientist who is proposing a solution to an as-yet-unsolved problem must proceed in this manner.

Other QUESTIONs will have definite answers, which are given at the end of this book. You can check these against your answers and gain confidence, or not, that you have the particular idea or technique asked about well in hand (and head). If so, you can proceed; if not, you must think some more about how to answer the QUESTION. The QUESTION below is one without a unique answer, although the hint should indicate a possible one.

• QUESTION 1.1.   Although the demonstrations support the model of the physical nature of sound, that it is a series of high and low pressure regions passing through air, this evidence does not make the model correct or even unique. A different demonstration might contradict our model's predictions, or our demonstrations may also support the predictions of an entirely different model. This hostile tension between theory and experiment is a characteristic of science. Theories and models become true as more and more experiments verify them and, finally, people agree that they must be right.

Can you think of another model of sound that is in accordance with our demonstrations? If so, try to propose an experiment that would support your model over the one I've explained to you. These are not trivial tasks, so let me instead suggest a model that won't work and you can try to recognize why not.

The suggested model is that sound is a series of high speed, but invisible, particles created at the sound's source, some of which fly into your ear or the microphone. Argue why this particle model of sound is wrong.

Hint: You can hear around some corners and behind some screens. Particles usually fly in straight lines. •

Along with the assertions about sound in general I also claimed that musical sound has special properties. Certainly not everyone will agree whether or not a sound is musical. So much depends not only on the sound, but in addition, to the context in which it appears. I will not deal with these aspects, but will limit discussion of musical sound to the properties of the sound itself. This will lead to a restricted and conservative definition of musical sound, and will keep us away from aesthetical considerations, which are considerably beyond the scope of this book.

CHAPTER 2.
MUSICAL SOUND

2.1 What is Musical Sound?

What makes sound musical? All sound has the properties shown so far. Now we will begin to divide musical sound from all the rest, which I will call noise, and you will find out how to tell them apart. This division is arbitrary and somewhat outrageous, and for now, will even exclude from being musical sound some sounds made by a few musical instruments. But, it will fix your attention on some observable characteristics of the traveling highs and lows. As usual, a demonstration is useful.

DEMONSTRATION VIII: Musical Sound and Noise

Apparatus: Sound Sources, Microphone, Oscilloscope

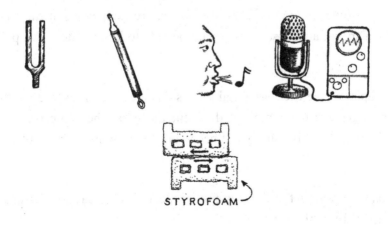

Part 1: Rules

A variety of sound makers are played. Some are musical; some are not. Their pressure patterns are observed on the scope's display.

The musical sounds show repeated patterns called cycles; noise does not have this repetition. Examples of a musical sound's scope displays look like this:

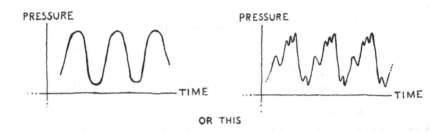

OR THIS

It is important to understand that this and the following pressure vs. time graphs are pictures of what's happening at some particular location, for example, at the entrance of the microphone. Later, for example, in Figure 2.1 you will see what's happening at many places at a single time. This is the same as a photograph.

In other words, when you see a graph of some changing quantity vs. time you must realize that these changes are happening at the same point in space; and, also, you should be able to know where this point is.

For graphs where the horizontal axis is distance, you must realize that the changes are happening at different places at the same time, i.e. simultaneously. It is usually not necessary to know just when this time was.

The basic pattern in the above graph is repeated. Whereas noise displays look more like this:

Chapter 2

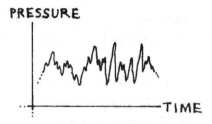

Listen to the recording of musical tones and noise (Band 2).

Thus, we are lead to the description:

> Musical sound is characterized by the reoccurrence of the same pressure fluctuation patterns: **cycles**.

You learned in Chapter 1 that the sound's traveling highs and lows hit the microphone and cause it to make electric signals as its diaphragm is slightly bent in and out. So, now that you know that the vertical axis on the scope display graphs represents pressure, you can draw them on ordinary pressure vs. time graphs. And, a little more detail can be added to them.

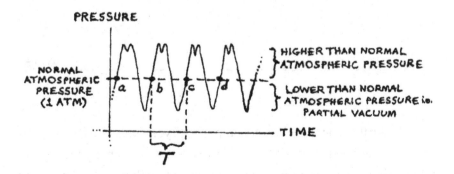

Part 1: Rules

New occurrences of the same pressure fluctuations begin at points a, b, c, d, ... A complete fluctuation is called a **cycle**, and a cycle begins at these points. I picked the starting of each cycle when the pressure was atmospheric and increasing, but it could have been started anywhere on the trace just as long as each subsequent cycle's start is at the same corresponding point. A cycle must always contain one complete fluctuation. Musical sound contains cycles.

• QUESTION 2.1.   Look again at the previous graph. Can you say for sure where on the pressure axis is the pressure zero, i.e. an absolute vacuum?

Hint: In Appendix A you will find that sound's high and low pressure fluctuations, above and below normal atmospheric pressure, are about only 1% changes from normal atmospheric pressure. If the above graph showed the zero pressure at the level of the time axis it would look like this:

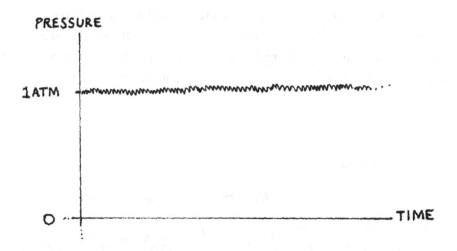

A graph like this does not show the important cyclic pressure fluctuations (they are graphed too small to recognize). So, it is often a good idea to graph only the pressure range near normal atmospheric pressure as the graph above this QUESTION does.

Chapter 2

Where on the pressure axis is the pressure representing an absolute vacuum? •

The horizontal axis on the above graph is time; further to the right is a later time. "Now" can be chosen to be at any point on this axis, but once it is picked, all of the axis to its right is the future, and all to its left is the past. Note, that each cycle takes the same interval of time. This is shown as T. This interval, T, is called the **period** of the sound, i.e., the number of seconds per cycle. A more common name used to describe sound is **frequency**. This is the number of cycles per second.

For example, if a sound wave has a period of (1/20) second, its frequency is 20 cycles/second. So,

$$T = 1/\text{frequency}.$$

Using f as the symbol for frequency we see that this equation is

$$T = 1/f. \qquad (2.1)$$

The previous demonstration showed the difference between musical sound and noise. You can play a tune with musical sounds but not noise. This could also be a defining characteristic of musical sounds: they have **loudness** and **pitch**. Let's see how these properties are represented on a scope's display.

DEMONSTRATION IX: Loudness and Pitch Displayed on an Oscilloscope

Apparatus: Audio Oscillator, Loudspeaker, Microphone, Oscilloscope

Part 1: Rules

First select a pitch from the audio oscillator-loudspeaker. Then notice the scope trace as you turn the loudness knob.

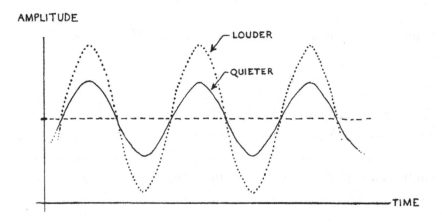

Listen to the recording of the changing loudness of a single pitched tone (Band 3).

You can see and hear that the amplitude of the waveform (the amount of maximum pressure change from atmospheric pressure) is a measure of the sound's loudness. It is not correct to say that amplitude equals loudness, because your ears do not hear tones with the same amplitudes but different frequencies as equally loud. Your ears are not equally

Chapter 2

sensitive to all frequencies. Of course not; some frequencies you cannot hear at all.

Now, notice the scope's trace as you change the pitch. Try to keep the loudness constant.

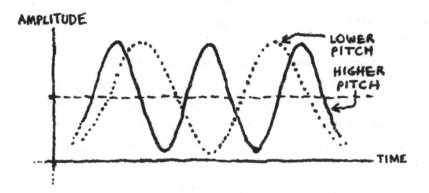

Lower pitch means bigger period and smaller frequency. Higher pitch is the opposite. Listen to the recording of tones with changing frequencies (Band 4).

Yet another term must be introduced: **wavelength**.

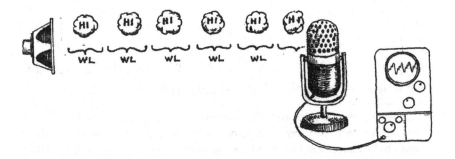

Part 1: Rules

The loudspeaker is producing high and low pressure regions as its cone vibrates while it broadcasts the sound. I've shown only the highs, in this case, one in each cycle. The wave moves with speed, v, toward the mike. One high is just hitting the mike and a period later the next will. If sound were visible the above sketch would be its photograph.

The distance a cycle occupies, measured in its direction of travel, is called a wavelength, WL. Five complete wavelengths are shown. Recall that the period is the time it takes for a cycle to occur. In this case, it is the time interval between arrivals of successive highs; so, it takes one period for the wave to travel one wavelength. Therefore, the words "frequency" and "period" refer to a cycle's repetitions in time (note that the horizontal axis in the previous graphs represents time), while "wavelength" is the length of a cycle in space. You know that

$$\text{Speed} = (\text{distance traveled})/(\text{time taken}),$$

therefore,

$$v = WL/T. \qquad (2.2)$$

v is expressed in units of meters/second, m/s. WL has the units of meters, m. T has units of seconds, s. And, the units are the same on both sides of the equal sign, as they must be in any equation.

Recall that

$$T = 1/f;$$

the frequency is the inverse of the period*. Thus, our previous equation can be written with T replaced by $1/f$. This produces,

---

* Because the period is measured is seconds, and the frequency is the period's inverse, the frequency has the units of 1/seconds, called "per second." Sometimes the frequency is specified as a number of "cycles per second." Adding the word "cycles" reminds you that there is a repetitive action occurring. Another word for "per second" is "hertz," abbreviated Hz. Heinrich Rudolph Hertz (1857-1894) made fundamental discoveries about electromagnetic

Chapter 2

$$v = (WL)\,f. \tag{2.3}$$

The speed of sound in air, v, is about 345 m/s and this is the value that will be used here. Because v is almost constant, the product of wavelength times frequency is also almost constant. Therefore, the above equation tells us that if the wavelength increases, the frequency must decrease, and vice versa. In particular, if the wavelength doubles, the frequency halves. In general, any fractional or percentage change in the wavelength will be accompanied by the same fractional or percentage change in the frequency in the opposite direction.

The speed-wavelength-frequency-period equations are true for all types of waves. This universality is the reason that the properties of sound waves can be shown using rubber tubing and a Slinky. It also allows you to observe the wavelength-frequency relationship directly and correctly but in another medium. Water waves have highs and lows, called crests and troughs, and a much slower speed.

DEMONSTRATION X: Traveling Waves Intersect, Superposition

Apparatus: Ripple Tank

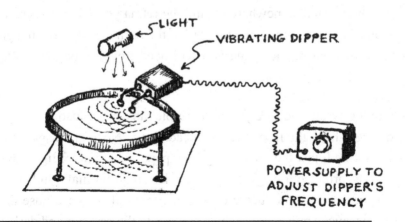

---

waves. So, a particular frequency might be written 200/s, or 200 cycles per second, or 200 Hz. All these variations are in common use, and this book will select the one that best fits the situation.

Part 1: Rules

The ripple tank is a shallow glass-bottomed pool of water, lighted from above, and with some type of dipping mechanism to make waves. The water in the waves' crests and troughs are lenses that focus or diffuse the light onto a sheet of paper underneath the pool to create patterns of bright and darker bands.

The glasses that you might be wearing are shaped into crests and troughs too.

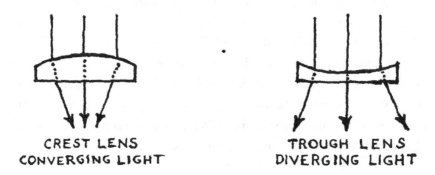

CREST LENS
CONVERGING LIGHT

TROUGH LENS
DIVERGING LIGHT

These lenses converge or diverge light rays, but probably not as much as shown above, to focus an image of what you are viewing onto the retina at the back of your eyeball. If your unassisted eye lens causes the image to be located somewhere behind the retina you will need crest-type glasses to move the image forward to the retina. And, if the image is located in front of the retina you will need trough-type glasses to move it back onto the retina.

The moving crests and troughs in the ripple tank water also converge and diverge the light from the overhead lamp. No real image of a scene is formed, just the locations on the paper where more light rays converge to produce a brighter spot or a less than normal number of rays hit the paper to produce a darker than normal spot. Because the crests and troughs move across the water, so do the brighter and darker bands on the paper.

Chapter 2

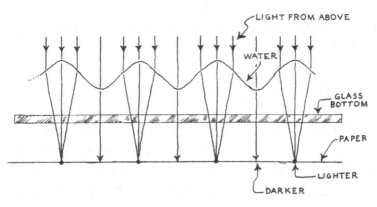

WATER CRESTS AND TROUGHS FOCUSING LIGHT

• QUESTION 2.2. Farsighted persons cannot see close objects clearly because the lenses of their eyes do not cause the light to converge and, thus, to focus on their retinas, but behind them instead. Therefore, they need the additional convergent focusing provided by glasses or contact lenses. Near sighted persons are the opposite, and need less convergent, or divergent, focusing.

In the above sketch is a water crest or a trough acting as a convergent lens?•

The waves move along the water's surface, and the bright and dark bands move across the paper. Figure 2.1 shows some of these bands. The brighter bands are shown as lines; the darker bands are in between them.

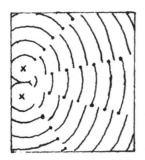 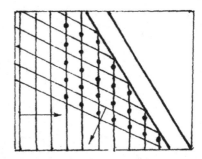

Figure 2.1 Ripple Tank Waves

- 27 -

Part 1: Rules

The left sketch shows waves moving to the right, away from the two dippers. The right sketch shows waves created by a single long dipper. They enter from the left and bounce off the barrier. Arrows show the directions of travel. In both sketches two waves cross. The dots indicate some of the locations where crests and troughs meet.

When there are two sources of waves, such as two dippers or when there is one source but its waves are reflected back on themselves by some obstruction, there are places where a crest from one wave passes through a trough from the other. The highs of the crests are diminished by the lows of the troughs and there is still water there. The value of the low is subtracted from the value of the high, and what's left is the resultant wave. Or, two crests could intersect producing a bigger crest. Their two values are added. Either of these is caused by a phenomenon called **interference** or **the superposition of waves**.

For musical sound, interference is absolutely necessary for the proper operation of instruments; otherwise it can be a disaster in some auditoriums, producing places where there is no, or too much, sound.

• QUESTION 2.3. What would you hear at the location where one sound wave's highs pass through another's lows? Where two high pressure regions pass through each other? At places where the interference always produces atmospheric pressure? •

Chapter 2

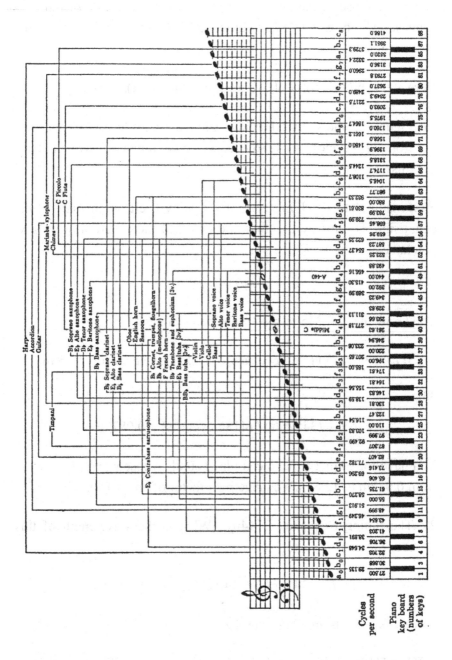

Figure 2.2 Pitch Ranges of Voices and Musical Instruments

Part 1: Rules

The numbers of the black keys are not shown, but are obvious. Their letters are those of the nearest white keys with sharp (#) or flat ($\flat$) signs added. For example key number 36 is $G_3\#$ or $A_3 \flat$. Note that the numerical subscripts change on C, not on A.

Here are some examples of using the equation (2.3), $v = (WL)f$.

1) Figure 2.2 shows that the piano can play notes having frequencies from 27.6 cycles/s to 4186 cycles/s. What range of wavelengths is this?

First manipulate $v = (WL)f$ so that WL is by itself on one side of the equal sign. Do this by dividing both sides of the equation by f. Now you have,

$$WL = v/f.$$

You know that $v = 345$ m/s, so plug in the values of f:

$$WL = (345 \text{ m/s})/(27.6 \text{ cycles/s}) = 12.5 \text{ m/cycle, or just } 12.5 \text{ m,}$$

and

$$WL = (345 \text{ m/s}) / (4186 \text{ cycles/s}) = 0.082 \text{ m}.$$

You have found that a piano can play sounds whose wavelengths vary from 0.082m to 12.5 m, a change by a factor of about 150.

2) The tuning frequency is 440/s. What is the wavelength of this tone?

$$WL = (345 \text{ m/s}) / (440 \text{ cycles/s}) = 0.78 \text{ m}$$

Some instruments are about this size.

• QUESTION 2.4. A microphone is attached to an oscilloscope and this wave is displayed.

Chapter 2

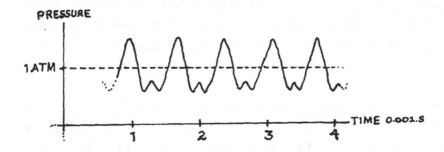

Identify some times when the pressure is atmospheric, greater than atmospheric, and a partial vacuum.

Identify some times when the concentration of air molecules next to the mike was less, the same, and bigger than in the normal atmosphere.

Is this sound noise? Why?

What is the sound's period? Its frequency? Its wavelength? •

2.2 Pitch and Frequency

The language of musical sound has its own vocabulary, perhaps more in common use than the technical-scientific one we've been using so far. In either case, continuing to use only the latter is something like trying to explain the wetness of water using only the language of molecular bonds. It can be done, but not in a friendly way. Of course, by now you have a working knowledge of some of the tech-sci vocabulary, but let's not neglect the musical one.

So far, the pitch of a sound has been specified by giving its frequency; this is the scientific way. The musical way to show pitches is to put them as notes on a stack of five horizontal lines and the spaces between them, called a **staff** (two are shown in Figure 2.2) and to give each note the name of a letter of the alphabet, A through G. Lower case letters are sometimes used. The musical interval between any lettered note and

Part 1: Rules

the next one above or below it with the same letter is called an **octave**. Figure 2.2 shows the ranges of pitches for some musical instruments using this scheme. There is no note with a frequency of 100/s, but I often will use this frequency because it makes calculations easier.

The note on the line between the top and bottom staffs is called C_4 or middle C, and the other lines and spaces have their letters in alphabetical order. The third from the bottom space in the top staff is also a C, C_5, and you can see how the subscript indicates which C-through-C octave you're in. For example, an all white note octave from D_4 to D_5 has these notes: $D_4 E_4 F_4 G_4 A_4 B_4 C_5 D_5$. The subscripted number changes on the letter C, not A. Listen to them on the recording (Band 5), and note that they are not all the usual notes in an ordinary scale starting on D. Two of them have too low a pitch.

Although audio oscillators allow us to produce any frequency tone we want, not all of these will have names, or be notes on staffs such as shown in Figure 2.2. The length of the piano keyboard and the naming of its keys have fixed the way notes are designated in octaves. Here are these notes, without subscripts, shown on a piano keyboard.

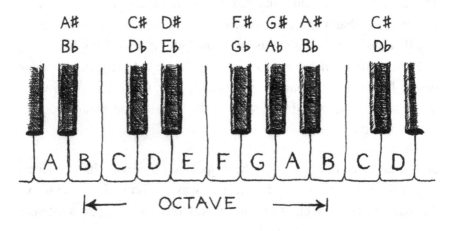

Figure 2.3 Piano Keys

Chapter 2

You can start on any white key and seven white keys later (or earlier) you will be at the same lettered key. This, too, is an example of an octave.

What about the black keys? For reasons partly historic and partly scientific, the octave has been divided in to 12 intervals, and this needed 13 keys. The notes in an octave divided this way comprise what is called a **chromatic scale**. The eight tones named below comprise what is called simply a **scale**. The black keys were added between the eight white ones as shown above. Now, going from one key to the next, white-to-white, white-to-black, or black-to-white, raises the pitch by an amount called a **semitone**. The black keys do not have separate letters. Sharps, ♯, or flats, ♭, are added to show pitches a semitone higher, or lower, than the letter. For example, C♯, G♯, and B♯ are shown on the keyboard. C♯ is now the same as D♭, but at one time they had slightly different frequencies. Note that B and C, and E and F do not have black keys between them. They have already been given frequencies a semitone apart. C and D, D and E, F and G, G and A, and A and B are two semitones, or a whole tone apart. The chromatic scaled octave contains 13 keys, each a semitone apart. Convince yourself of this by counting the number of keys in any octave starting and ending on the same named key.

When you first learned music you might have been taught these names for the notes in a scale: do, re, mi, fa, sol, la, ti, do. Also eight notes, and this is no coincidence. A scale is also the notes that span an octave. For example, if do is a particular lettered note, such as C♯, the next do an octave higher, or lower, will also be a C♯. Subscripts identify the particular C♯, a do at C_5♯ will have the next higher pitched do at C_6♯ and the next lower pitched one at C_4♯.

The term "mode" has two distinct meanings. You will see it used to identify the harmonics of a vibration, the scientific meaning. Mode also has a musical meaning. At one time there were several schemes for determining the number of semitone intervals between the notes in a scale. Each of these schemes was given a name and as a group they were called the various modes of producing a scale. Today only two of

Part 1: Rules

these modes are in general use: the Major mode and the Minor mode. The Chart 2.1 below shows how many semitones separate the notes in Major and Minor mode's scales.

	do	re	mi	fa	sol	la	ti	do
Major	2	2	1	2	2	2	1	
Minor	2	1	2	2	1	2	2	

Chart 2.1, Semitone Intervals in Major and Minor mode Scales

Note, that both the Major and Minor mode's chromatic scales contain 12 semitone intervals, as they must, since they both span an octave interval. You can start a Major or Minor mode scale on any note if you keep these intervals between the notes. The piano key you use for do and the name of the mode will be the **key signature**, or **key** of the scale. Using the keyboard sketch above you can identify the C Major scale as C D E F G A B C. The C minor scale is C D D♯ F G G♯ A♯ C. The word "Major" is often omitted in Major mode key signatures.

In addition to being a piano key, each semitone in a Major chromatic scale is given a name: do, di, re, ri, mi, fa, fi, sol, si, la, li, ti, do. These names are used when the scale is sung. This naming scheme does not work for a Minor mode chromatic scale. For example, Chart 2.1 shows the Minor mode's re and mi are a semitone apart and there is no "room" for a ri. Similarly, sol and la are only a semitone apart and there is no "room" for a si. Also in the Major mode ti and do are a semitone apart, but two semitones apart in the Minor mode. There is no name for this extra semitone.

• QUESTION 2.5. What are the piano keys in the G Major scale? What are the piano keys in the G♯ Major scale? What are the piano keys in the G♯ Minor scale?

Chapter 2

Locate a piano or keyboard and play, and hear, these scales. This will be much better than listening to them from a recording. •

• QUESTION 2.6. You were previously shown the following notes: D E F G A B C D. They do not form a Major scale. Why? Find and name the two incorrect notes and state what they would be in a D Major scale. •

• QUESTION 2.7. Do the notes D E F G A B C D constitute a D Minor scale? Why? What note change(s) must be made to make the above notes into a D Minor scale? •

You've seen from the above questions that the D to D white note scale is neither Major nor Minor mode. It is the Dorian (also called Doric) mode.

NAME	SYMBOL	INTERVAL FROM do	SEMITONES IN INTERVAL	NOTE SPAN FROM	RATIO OF FREQUENCIES
UNISON		do-do	0	C-C	1:1=1.000
MAJOR THIRD	M3	do-mi	4	C-E	4:5=0.800
PERFECT FOURTH	P4	do-fa	5	C-F	3:4=0.750
PERFECT FIFTH	P5	do-sol	7	C-G	2:3=0.667
MAJOR SIXTH	M6	do-la	9	C-A	3:5=0.600
MINOR THIRD	m3	do-ri	3	C-E♭	5:6=0.833
MINOR SIXTH	m6	do-si	8	C-A♭	5:8=0.625
OCTAVE		do-do'	12	C-C'	1:2=0.500

Table 2.1

Musical Intervals, Just Tempering, Major Mode. The Prime Sign Indicates the Next Higher Octave

Part 1: Rules

There are other important musical intervals beside the OCTAVE and semitone. Table 2.1 names the more common ones and shows their symbols and other information about them. The most melodic ones were once called the THIRD, FOURTH, and FIFTH, but these names have been changed to MAJOR THIRD, PERFECT FOURTH, and PERFECT FIFTH. These numbered names once indicated the number of notes in the interval, but now are just the number of white keys in the interval if the first note is a C. Listen to these intervals from the recording (Band 6). The number of semitones in a Major mode interval shown in Table 2.1 is correct for an interval between any two notes. The note names are applicable to Major mode scales only; in the Minor mode there is only one semitone between sol and la, and there is no room for a si. Referring to Table 2.1 you can also see that the MINOR SIXTH interval in a Minor mode key would be the 8 semitones between do and la. Similarly the MINOR THIRD interval for a Minor mode key would be the three semitones between do and mi. Nevertheless these names and symbols are still used. From now on only the properties of the Major mode will be used in this book unless noted otherwise.

The scientifically interesting feature of the intervals is that the ratios of the two frequencies are the ratios of integers, and rather small integers, too. However, if you compare the ratios shown in Table 2.1 to the ratios you can calculate from the actually used frequencies shown in Figure 2.2 you will find that they are not quite the same. They are within 3% of each other, and this small difference is the result of a very important decision made about 400 years ago, and which will be explained in Chapter 4 when you discover the problems and solutions of making musical instruments able to play in all key signatures. For now, I'll just say that there are several schemes for specifying the frequencies of the lettered notes; the one that results in the intervals being exact ratios of integers, such as shown in Table 2.1 is called "just tempering." The frequencies of the notes given in Figure 2.2 are the result of another called "even tempering," and which will be introduced in Chapter 4.

• QUESTION 2.8. Here are three frequencies of sound waves.

Chapter 2

440/s

220/s

880/s

What are the intervals between these tones?

If you increase the pitch of a tone by two octaves what happens to its wavelength? •

• QUESTION 2.9. You play the tone do on your guitar with a frequency of 345 Hz, and a friend plays la in the same scale on his flute. Assume just tempering.

What is the name of the interval?

What is the frequency of la? •

• QUESTION 2.10. You play a note you call do on your violin. Its frequency is 700 Hz. What are the frequencies of the first mi, fa, and sol *below* your do? Assume just tempering. •

• QUESTION 2.11. Calculate the frequencies ratios of three PERFECT FIFTHs picking the frequencies from those shown in Figure 2.2. Pick your PERFECT FIFTHs from low, mid, and high pitches. Compare your ratios to the just tempered ratios shown in Table 2.1.

Do the same thing for three MAJOR THIRDs.

Do the same thing for three PERFECT FOURTHs.

Which interval, PERFECT FIFTH, MAJOR THIRD, or PERFECT FOURTH, gives the best agreement between just tempering and the ones you used? Which gives the worst? •

CHAPTER 3.
MUSICAL VIBRATIONS AND THEIR VIBRATORS

Here begins your look at vibrating objects that can produce musical sound. The tuning fork is one, but is restricted to playing a single pitch and is not much of a tune maker. The electronic audio oscillators are others, but they also are not the kind of music makers we will examine. Remarkably, this narrows down the field to just two kinds of vibrators: the flexible string, either the normal thin shape or one so much thicker that perhaps the name string is not applicable, and the elastic air within a musical instrument.

3.1 Vibrating Strings

The thin strings are simpler to explain; and we'll start with them. Our example is a long rubber cord whose vibrations are big and slow enough to be easily observed and measured.

DEMONSTRATION XI: Standing Waves

Apparatus: Rubber Tubing

The tubing is fixed at both ends between walls a distance L apart. Start the tube vibrating up and down by shaking it lightly close to one of the walls. Experiment with a range of shaking frequencies. At certain vibrational frequencies the tubing takes on shapes called **standing waves**.

Part 1: Rules

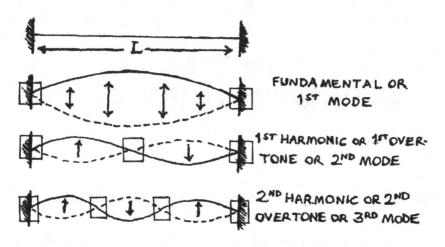

Although your shakes are producing deformations that are traveling down the tube just like you've seen before, when the standing waves appear there is no longer any indication that this is happening. The tube just moves up and down, and some parts of it don't move at all. The places along the tube where there is no motion are called **nodes**, and where the motions have the same up and down amplitude are called **antinodes**. The small squares show the locations of the nodes of the first three modes. The locations of the antinodes are midway between the nodes.

Based on these sketches you should be able to draw the standing waves for any mode.

When standing waves occur the string is "playing" one of its **modes** of vibration. The lowest frequency mode is called the **fundamental mode** or just the **fundamental**. The fundamental mode is also called the 1st mode of vibration of our string. The higher modes of vibration are numbered 2nd, 3rd, etc. If, in addition, the frequencies of the higher modes are integer multiples of the frequency of the fundamental (or 1st) mode, these frequencies are called **harmonics** and are numbered 1st harmonic, 2nd harmonic, etc. Unfortunately, this means that the 1st harmonic has the frequency of the 2nd mode: the numbers don't

Chapter 3

match. Here is a chart showing this confusing nomenclature. Each vertical column contains the alternative names for the same vibration of a harmonic oscillator.

MODE	Fundamental 1st Mode	2nd Mode	3rd Mode	...
HARMONIC	Fundamental	1st Harmonic	2nd Harmonic	...

You found that you had to shake the tubing faster and faster to produce each higher mode. Musicians can "keep time," and can note that the frequencies of the first three modes had frequency ratios $f_1 : f_2 : f_3 = 1 : 2 : 3$. The subscripts will be used to identify the mode. But beware, in Section 7.4.2 of Chapter 7 you will learn that some wind instruments have $f_1 : f_2 : f_3 = 1 : 3 : 5$ ... The notation $f_1 : f_2 : f_3 = 1 : 2 : 3$ is shorthand for "the frequency of mode 1 is to the frequency of mode 2 as 1 is to 2; and the frequency of mode 2 is to the frequency of mode 3 as 2 is to 3." So, the colon is read, "is to" and the equal sign is read, "as." So, $f_1 : f_2 = 1 : 2$ can also be written as,

$$f_1 / f_2 = \tfrac{1}{2}.$$

Although a wavelength is the length of a cycle, it's not always easy to recognize one, especially in standing waves. In the sketches of the standing waves, the wavelength of the 1st mode is $2L$; the 2nd mode's wavelength is L; and the 3rd mode's is $(2/3)L$. Do you see the rule for finding these wavelengths? Here it is.

A cycle must

- be the length of a repetitive part of the wave

- have the slope of its shape continuous at its boundaries.

In the 1st mode if you attach another standing wave to the one in the sketch there would be a kink in the shape. This violates the second

Part 1: Rules

requirement. You have to imagine the phantom missing part of the 1st mode, which occupies another length L.

You can attach another 2nd mode's standing wave to the one shown and you will get both another repetitive part and the slopes that are the same at the join. This meets both requirements and so the wavelength is L.

The sketch of the 3rd mode shows one and one half wavelengths.

In DEMONSTRATION V you saw a pulse of deformation travel down a length of rubber tubing. You might have noticed that the pulse bounced back from the tube's or Slinky's clamped end. At first glance this has nothing to do with the locations of the nodes and antinodes of standing waves, which, as you have just seen in DEMONSTRATION XI, do not move along the tube at all. Your first glance is wrong. As you will discover later, standing waves are the result of the combination and addition of the amplitudes of waves having the same frequencies passing through each other as they cross. In the stringed and most wind instruments this is accomplished when the wave through the string or the air inside the instrument and its reflection cross, traveling in opposite directions. Because both of these waves have the same origin, the necessary condition that they have the same frequencies is automatically satisfied.

Because a musical tone has cycles that repeat in time, it also has wavelengths that repeat in space. You can examine the spatial structure of the wave's modes and identify the cycles that are their wavelengths. Then putting this information into equation (2.3),

$$v = (WL) f,$$

will allow you to solve for the frequencies. This is another way to find out that they are integer multiples of the 1st mode's frequency. Here's how.

First recall that a cycle of a simple wave is shown on a pressure vs. time graph as follows.

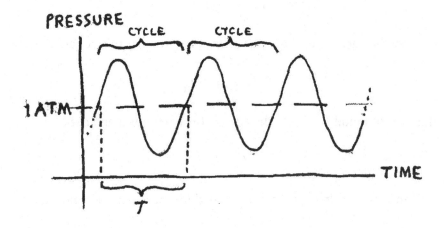

The high and low pressure regions are also spread out in the space between the wave's source and the sound detector, and a "photograph" of this would make a graph like this:

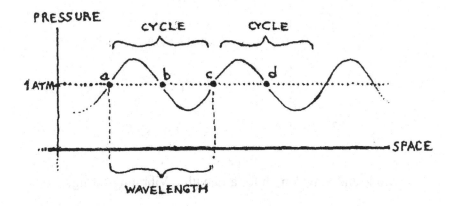

Part 1: Rules

The horizontal axis is no longer time. This graph is what's happening at the same time at various places along the wave's path. Now compare this graph to the sketches of the first three mode's standing waves in DEMONSTRATION XI. You can do this; they are all photos of waves. For the 1st (or fundamental) mode, note, that the length L is also the distance between points a and b. So,

$$L = (1/2)\ (WL \text{ of the 1st mode}).$$

I'll use the subscript to identify the mode and write this equation as,

$$L = (1/2)\ (WL_1).$$

For the 2nd mode, L is the distance between points a and c. So,

$$L = WL_2.$$

L is the distance between points a and d for the 3rd mode, and so,

$$L = (3/2)\ (WL_3).$$

Collecting these results in a table:

MODE	WAVELENGTH FORMULA
1	$L = \tfrac{1}{2}(WL_1)$
2	$L = \tfrac{2}{2}(WL_2)$
3	$L = \tfrac{3}{2}(WL_3)$

Table 3.1
Mode and Wavelength for a Length L Vibrating String.

Chapter 3

Examine the columns containing the formulas for the wavelengths in Table 3.1, and recognize that all the formulas can be written as a single equation:

$$L = (n/2)(WL_n), \qquad \text{with } n = 1, 2, 3,$$

where n is the mode number. This formula is general and works for all the modes. Do the algebra to rearrange the terms and find the general equation for some mode's wavelength,

$$WL_n = (2L)/n, \qquad \text{with } n = 1, 2, 3, \ldots \qquad (3.1)$$

Now you are ready to use the formula (2.3),

$$v = (WL_n) f_n$$

to find the frequencies. Actually not; remember that this is an analysis of standing waves in a flexible tube. Here, the speed, v, used here is how fast the wave travels down the string or tubing and not the speed of sound in air. This speed depends on the construction of the string or tubing, and although you can combine

$$WL_n = (2L)/n$$

and

$$v = (WL_n) f_n$$

to find

$$f_n = v\, n/(2L), \qquad \text{with } n = 1, 2, 3, \ldots ; \qquad (3.2)$$

unless you know the speed, this formula is still incomplete. However, it does show that ratios of the modes' frequencies are ratios of the modes' integers. For example,

$$f_2 / f_3 = 2/3.$$

Part 1: Rules

Everything else cancels, and so,

$$f_1 : f_2 : f_3 = 1 : 2 : 3.$$

A general formula becomes apparent,

$$f_n = f_1(n), \text{ with } n = 1, 2, 3, \ldots \qquad (3.3)$$

The frequencies of the modes given by equation (3.3) are called the modes' **harmonics**. Equation (3.3) shows that these frequencies are integer multiples, 2, 3, 4, etc. of the frequency of the 1st mode.

• QUESTION 3.1. The frequency of the 1st mode of the tube shown in DEMONSTRATION XI was about 3/s. The length of the tube was about 3m. What is the speed of the wave traveling down the tube? Compare this speed to the average speed of the winner of a 100m dash. •

The wonderful thing is, that if you examine the frequencies that a string of length, L, can play, the musical intervals appear. Here is an example for a string whose 1st mode's frequency is 100/s.

Chapter 3

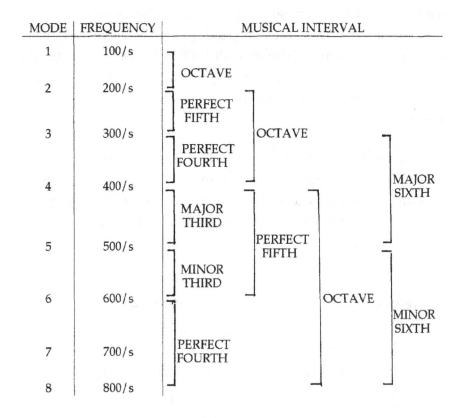

Table 3.2
Modes and their Musical Intervals

Compare the names and frequency ratios of the musical intervals in Table 2.1 with the frequencies of the modes of a string shown in Table 3.2, and note that the strings' modes' frequencies form musical intervals.

All the modes' frequencies, except the 7th mode's 700/s, are notes (do, mi, fa, or sol) on scales starting with do at 100/s. A stringed instrument of this sort, if you could persuade it to play the mode you wanted when you wanted, could play simple tunes such as bugle calls.

Table 3.2 shows the modes in the first three octaves. Note that each higher octave contains more of the notes in its scale. Also realize that if

Part 1: Rules

such a stringed instrument sounded several of its modes simultaneously they, except for the 7th, would be creating musical intervals and, thus, be musical all together. Here is another difference between a musical instrument and a noisemaker; and, as you will see in Chapter 5, many musical instruments use stretched strings as the source of their sound waves' frequencies.

Is it possible to influence, even to specify, the modes a string can play? It certainly is! It is this simultaneous mixture of modes that gives an instrument its unique sound, or **timbre**. An oscilloscope display having cycles with complicated shapes is an indication of several modes playing together.

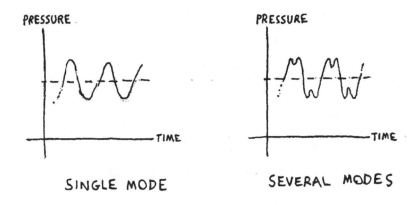

3.2 More Vibrating Strings, Preventing Modes, Combining Modes

The general technique for influencing the sounding of a mode in a string is to insert some device that forces a node (or antinodes) at some location along the string. If this forces a node (no motion) at a place where a mode's antinode would have to be, that mode is prevented. The name for this procedure is **clamping**. Here is an example.

Chapter 3

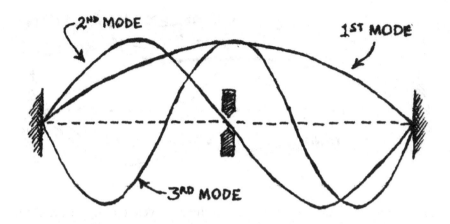

Here is the same tubing you saw in DEMONSTRATION XI with the first three modes shown together. Two small blocks have been placed halfway across the string at the position of the 2nd mode's central node. These blocks will clamp out any up or down motion at the midpoint of the string. This does not affect the 2nd mode; it already has a node there. But the 1st and 3rd modes need antinodes at the string's midpoint; and the blocks prevent their up and down motion. Thus, these blocks clamp out the 1st and 3rd modes. You should be able to see that these blocks will prevent all the odd integer modes and allow all the even integer ones. Stringed instrument players pluck or bow at various locations along the string to change the timbre of their instrument.

• QUESTION 3.2.   Where would you place the clamp if you wanted to prevent the first two modes but allow the 3rd mode? •

3.3 Complex Vibrations in a String

It's not easy to visualize the shape of a string's standing wave when it is vibrating with several modes at the same time. The resulting vibration is called a **complex vibration**. Let's draw one for a string fixed at both ends so we at least know that these ends are nodes. The question remains: are there any antinodes, and if so, where?

Part 1: Rules

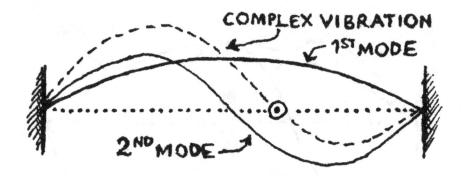

Here are the first two modes' standing waves (solid curves), the unstretched string's shape (dotted line), and the string's shape when it is playing both the modes together (dashed line). The sketch shows both modes at "full stretch" and so the string is at one of its maximum displacements, too. Of course, the string is vibrating up and down; this sketch is a snapshot photo at some time. The rule for finding the dashed line is to add the vertical displacements of each mode together, i.e., add the displacements of the solid curves. Here is another kind of the phenomena of interference and superposition that was introduced in Section 2.1 in Chapter 2, but here caused by the interference of oppositely traveling waves in strings.

If a displacement is below the unstretched shape, subtract it. At the place where the 1st mode's displacement is upward, and the 2nd mode's is downward by the same amount, they add to zero (the circled place), and if this cancellation continues at that place it is the location of a node.

But this is not the case here, because one half a period of the 1st mode later (which is one period later for the 2nd mode) the modes are again at "full stretch" but not in the same positions, and now the string and its modes look like this:

Chapter 3

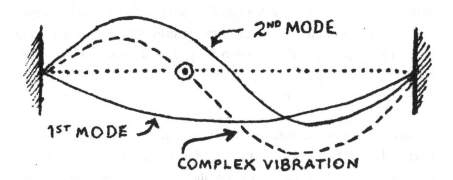

Compare the two sketches and see that the place along the string where it has its greatest displacement has moved. In the first sketch it was on the left; in this one it had moved over to the right. These maxima are not locations of antinodes. They do not have equal up and down displacements, and they are not midway between the nodes as you saw necessary in DEMONSTRATION XI.

Does the up and down motion of the center of the string meet the requirements for an antinode to be there? Yes, this motion takes place half way between the nodes and has equal amplitude ups and downs. Nowhere else along the string are these conditions satisfied.

The string's changing shape from time to time is a sort of back and forth sloshing motion, and not easy to visualize. This is why DEMONSTRATION XI shows only one mode at a time. You will see later that the complex vibrations of the whole musical instrument itself produces sound waves, which in turn contain several overtones, and thus, are also complicated.

3.4 Combinations of Musical Sound Waves

The number and amplitude of the modes of vibration of a musical instrument's complex vibrations produce the various frequency sound waves that give the instrument its tone and timbre. The resulting sound wave is also a **complex wave**, and the process of finding its shape by adding or subtracting the amplitudes of intersecting traveling waves is also **superposition**. The word "superposition" is also used for the name of a complex wave. In Section 3.3 you used superposition to find the shape of a vibrating string. Here it will be used to find the shape of the pressure vs. time sound wave graphs.

Complex waves can be generated by a single musical instrument, as just described, or by the intersection of sound waves from different sources. Microphones and ears are important places where these intersections can occur. Any number of traveling waves can combine; imagine the complex wave created by a whole orchestra. But, examining just two intersecting single frequency sound waves is enough to show the main features of complex waves. More than two only add complications, not new ideas.

The pressure changes at the point where two traveling sound waves intersect. Two high pressure regions combine to create a higher pressure region; or a high and a low (partial vacuum) could combine and cancel each other (creating atmospheric pressure).

Here is a pressure vs. time graph showing one cycle of what happens where the fundamental and 1st harmonic of some wave are present. They are drawn "in phase." This means that they are starting from atmospheric pressure at the same time, and both are displacing toward higher pressure.

Chapter 3

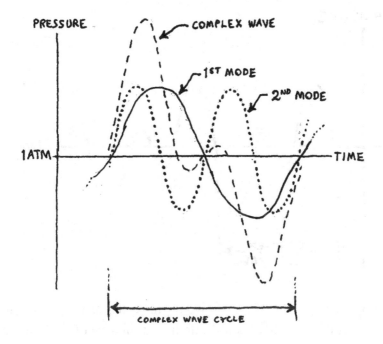

From equation (3.3) you can find out that the frequency of the 1st harmonic is twice that of the fundamental: $f_2 = 2f_1$. Use equation (2.1) to replace the symbols for the frequencies with the symbols for their periods, i.e., f_1 and f_2 with T_1 and T_2. Now you will have an equation: $T_1 = 2T_2$. Thus, in the time it takes the fundamental to complete one cycle, the 1st harmonic completes two cycles, as the above graph shows.

The location of this intersection might be at a microphone, and an oscilloscope display would show only the complete complex wave. The above graph presents more information; it is drawn showing both modes and the resulting complex wave. Only one cycle of the 1st mode and two cycles of the 2nd mode are shown. This is enough to show one complete cycle of the complex wave.

The whole oscilloscope display would look like this:

Part 1: Rules

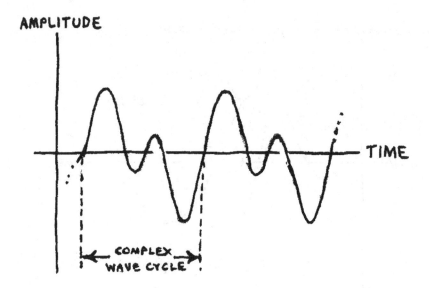

The complex wave, in this case, would have the same period, and frequency, as the 1st mode. You would hear a mixture of two tones an octave apart, i.e. the OCTAVE interval. This complex wave has been made for the case where the 1st and 2nd modes have equal amplitude pressure fluctuations and you should have no trouble now sketching the complex wave where the modes' amplitudes are not the same height.

It is also possible to play these two waves "completely out of phase." This means that as the 1st mode is starting upward the 2nd mode is starting downward. The frequencies themselves do not change, only how they begin relative to each other. Here is a sketch of the first two modes completely out of phase.

Chapter 3

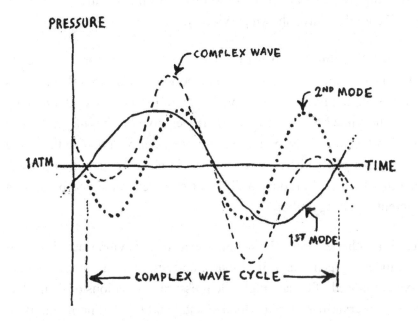

And the whole display looks like this.

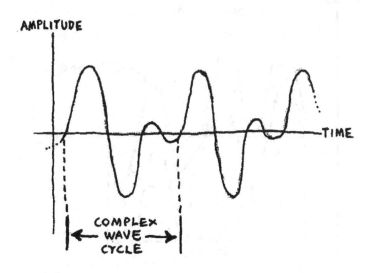

This complex wave above looks similar, but not identical to the "in phase" complex wave shown previously.

Note that the complex waves from both the "in phase" and "completely out of phase" conditions have the same periods. This correctly indicates that they should have the same pitch, that of the 1st mode. But do they sound the same? No, their timbres are perceived to be different[2]. The situation becomes even more complicated when you realize that there is a range of phase differences between "in phase" and "out of phase," each of which will produce a different complex wave and possibly a different resulting timbre.

Here is another example. Now, suppose a string is vibrating with equal amplitudes of its 1st and 3rd modes, and produces a complex sound wave composed of equal amplitude pressure fluctuations of them. The musical interval between the fundamental and the 2nd harmonic is an octave plus a perfect fifth (see Table 3.2).

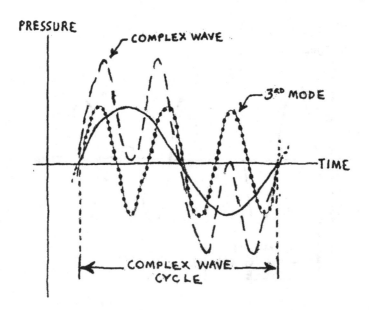

Chapter 3

Note that, in this case, the complex wave's period is the same as the 1st mode's.

The oscilloscope would show:

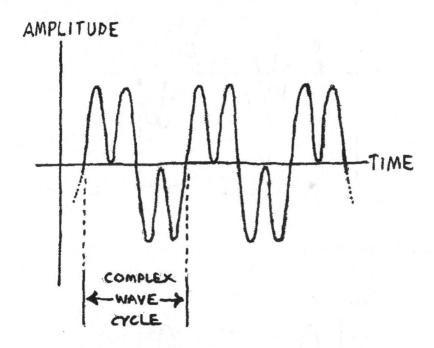

Here, as in the previous example, the complex wave has the same period as the 1st mode. So, in both these examples you will hear the same pitch but different timbres. Listen to the recording of the sounds for this and the previous example (Band 7).

• QUESTION 3.3.   Sketch the complex wave for a superposition of the 1st and 3rd modes completely out of phase with each other. Compare your sketch to the in phase example. •

Here is one more example, and one in which something new appears. The string is vibrating with equal amounts of its 2nd and 3rd modes. You expect to hear an interval of a perfect fifth, and you do. But,

Part 1: Rules

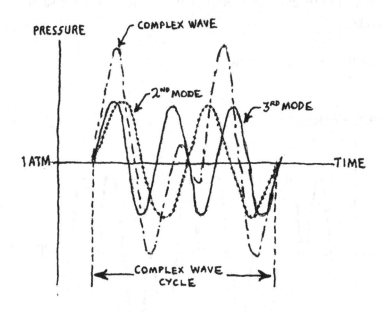

And the scope displays:

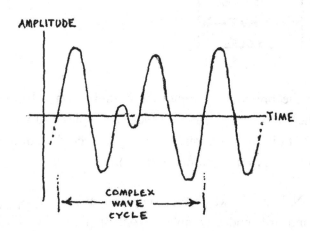

Now the period of the complex wave is neither that of the 2nd nor the 3rd mode. The sketch shows that its period is twice the 2nd mode's and three times the 3rd mode's. A new tone has sounded. It is the 1st mode of this harmonic series, and your ear is probably fooled into believing

Chapter 3

that the pitch of this complex tone is that of the 1st mode. You will hear three tones and they will produce intervals of an OCTAVE (1 : 2) and a PERFECT FIFTH (2 : 3). It is musical sound but the harmonies are richer than you might have expected. Listen to the recording of these two pitches being played (Band 8).

• QUESTION 3.4.   The string is vibrating with its 3rd and 4th modes, and you hear the resulting sound. Sketch and identify a cycle of the complex wave as seen on an oscilloscope's display. Identify the musical intervals present. Listen to the recording of this interval (Band 9). •

In practice the given tones are not equally loud; the higher harmonics usually have less amplitude.

3.5 Beats

The complex wave for two tones having almost the same frequencies has special characteristics which when heard are called **beats**. Instead of hearing other pitches you hear the sound become alternatively louder and softer; these are the beats. Listen to the recording of this (Band 10).

The explanation of how beats occur is made with the same graphical addition of displacements you've just seen, but because the two tones are almost in unison it is almost impossible to draw the complete graphs. They will be too long to fit on a page. Nevertheless, let's start graphing two waves having almost the same periods and starting in phase; and show only parts of the graph. For simplicity make the two waves have equal amplitudes.

Part 1: Rules

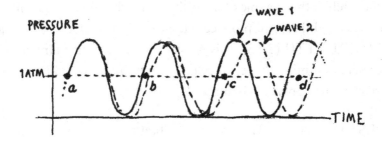

The waves are starting a cycle together at time a. By the time they begin their next cycles, near point b, wave 1, with a smaller period, begins before wave 2. Similarly, when they begin their third cycles, near point c, wave 1 starts even earlier relative to wave 2. As this continues, soon wave 1 will be starting its cycle when wave 2 is only half way through its cycle. The graph will then look like this:

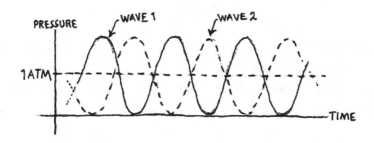

They are almost completely out of phase now, and the displacements add almost to zero. One is positive and the other is about the same amount negative. When they first began they were both positive, or negative, together, and the summed displacements added to give a big amplitude. This process of wave 1 beginning its cycle progressively earlier relative to wave 2 continues, and later both waves will be starting a cycle together again. Not the same cycle though; if, for example, wave 1 is beginning its 100th cycle, wave 2 will be beginning its 99th.

The first graph showed that the amplitude of the complex wave was big; in the second graph the amplitude was almost zero. A sketch of the

Chapter 3

complex wave on a shrunken time scale (many more seconds shown) axis would look like this:

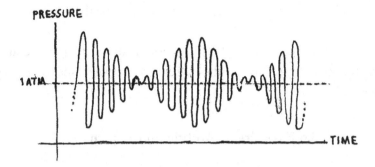

You hear a tone about the same pitch as the original waves', but the loudness of this tone slowly increases and decreases. These loud-soft cycles are the beats, and their slow cycling takes place at the **beat frequency**. If the waves had exactly the same frequencies there would be no beats, i.e., a zero beat frequency. Musicians get in tune by playing the same note and adjusting their pitches until the beats disappear.

As the frequencies of waves 1 and 2 gradually get increasingly different, the beat frequency increases. It goes from being a loud-soft variation, to kind of a rough chatter, and then, when it reaches a frequency of about 50 Hz becomes a tone itself. Listen to the recording of this happening (Band 11).

The period of the beats is the shortest time interval during which wave 1 will make exactly one more cycle than wave 2. Here is an example of applying this rule.

You are hearing two tones: 402/s and 400/s. At the end of 1 second wave 1 is beginning its 403rd cycle and wave 2 is beginning its 401st. So, in 1 second wave 1 has made exactly two more oscillations. In what time interval will only one more oscillation be completed? Of course, in 1/2 second. The beat period is 1/2 second and the beat frequency is 2/s. Listen to the recording of these beats (Band 12).

Part 1: Rules

How about if the frequencies are 300/s and 305/s? After some thought you realize that in 1/5 second one wave has oscillated 60 times and the other 61 times. Now the beat period is 1/5 second and the beat frequency is 5/s. Listen to the recording of these beats (Band 13).

A rule that usually works if the tones are close together in frequency is that the beat frequency is the difference between the two waves' frequencies.

3.6 Ears Create Additional Frequencies

The ultimate detector of music is the ear. Microphones and oscilloscopes have their place, but a scope display is not music. The individuality of persons extends to what they hear. This is not just a difference in perceived aesthetic content of the music, but includes the actual tones. The ear creates tones; the amount differs from person to person.

The ear is, among other things, an amplifier. Perhaps a transducer (a name introduced in DEMONSTRATION II in Chapter 1) would be a better word. It changes the pressure fluctuations in the air into electrical-chemical changes in your nervous system, which your brain finally tells you are sound. How well does it do this? Is your ear-brain a high fidelity device? Not very. Your ear-brain, in common with any nonlinear amplifier, creates frequencies. "Nonlinear" means that the amount of amplification depends on the frequency and amplitude of the input signal. Most real sound waves are complex waves made of very many single frequency waves. You will learn more about this in Appendix B. Your nonlinear ear-brain system does not process all frequencies equally; it creates a modified complex wave for you to "hear." This in turn means that this modified complex wave is made from a different set of single frequency waves. Most of these waves are the same as were in the original complex sound wave, but not all are. The frequencies of the newly created waves are **called combination frequencies**, and they do not exist in the sound waves in the air, only in your ear-brain system. The genesis of these frequencies can be found in the mathematical formulas describing nonlinear systems. The math tools needed to produce and manipulate them are not included here, but the results can be. They are as follows: if two tones with frequencies $f1$ and $f2$ enter the ear, the following combination frequencies will be generated,

$$2f1, \ 2f2, \ f1+f2, \ f1-f2$$

These combination frequencies are not as loud as the real sounds that entered the ears, and how much of each one is present depends on the person's hearing. Every person hears these combination frequencies along with the real tones, the complex frequencies, and the beats; and thus, everyone perceives a slightly different sound. Perhaps this is why we can't agree on the music's sound, especially when trying to describe a room's acoustics.

3.7 Analysis of Two Sounds

Two tones are produced somewhere and your ears intercept them. A complete analysis of "what you hear" must include the tones, the complex wave, beats, and the combination frequencies. Given the two tones, you have seen how to determine the rest. Here are some examples.

Example 1.)   A PERFECT FIFTH consisting of 200 Hz and 300 Hz are played. What do you hear?

The complex wave is gotten from a graphical construction.

Part 1: Rules

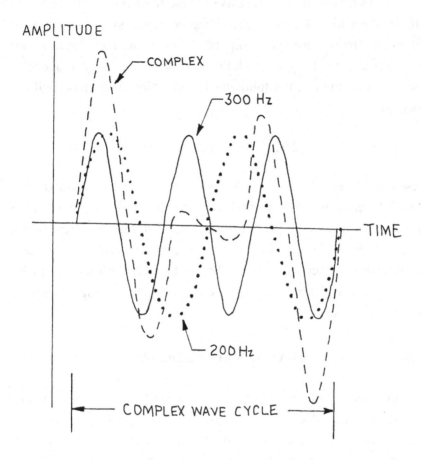

One cycle of the complex wave is drawn. Its period is three times as long as that of the 300 Hz wave and twice as long as that of the 200 Hz wave. Thus, its period is

$$3(1/300) \text{ s and/or } 2(1/200) \text{ s},$$

which is 1/100 second; and its frequency is 100 Hz.

Determining the beats and the beat frequency is more complicated. Applying the rule that usually works (the difference of the frequencies) produces a beat frequency of 100 Hz. Is this a beat? No. 100 Hz would be perceived as a tone, and not loud-soft fluctuations.

The combination frequencies are straightforward calculations.

FORMULA	COMBINATION FREQUENCY
$2f1$	2(300 Hz) = 600 Hz
$2f2$	2(200 Hz) = 400 Hz
$f1 + f2$	300 Hz + 200 Hz = 500 Hz
$f1 - f2$	300 Hz - 200 Hz = 100 Hz

Put all this into a table and complete the analysis by identifying the musical intervals.

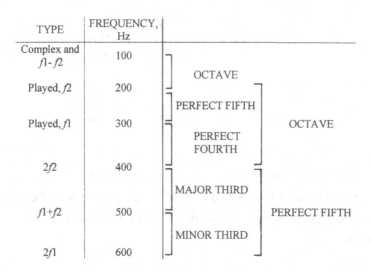

You hear the frequencies of the first six modes (fundamental plus five harmonics) of a harmonic series built on the 100 Hz tone. Furthermore, you hear musical intervals in addition to the PERFECT FIFTH played. Our analysis has discovered a sound with a timbre constructed from this collection of musical intervals. Listen to the recording of the perfect fifth again (Band 8).

Part 1: Rules

Example 2.)   A PERFECT FOURTH interval consisting of 300 Hz and 400 Hz tones is played. What do you hear?

A graph shows the complex wave.

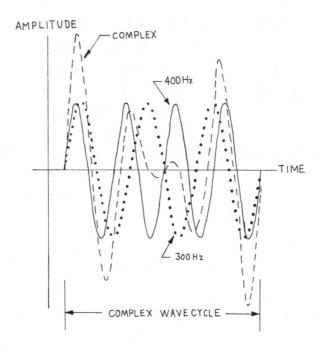

One cycle of the complex wave is drawn; and in this case its period is again 1/100 second and its frequency 100 Hz. There are no beats for the same reason explained in Example 1.). The table of the combination frequencies is

FORMULA	COMBINATION FREQUENCY
$2f1$	800 Hz
$2f2$	600 Hz
$f1 + f2$	700 Hz
$f1 - f2$	100 Hz

Chapter 3

Putting this together as before and indicating the musical intervals produces

TYPE	FREQUENCY, Hz	MUSICAL INTERVAL
Complex and $f1-f2$	100	
Played, $f2$	300	⎤
		⎥ PERFECT FOURTH
Played, $f1$	400	⎦ ⎤
		⎥ PERFECT FIFTH ⎤
$2f2$	600	⎦ ⎥
		⎤ ⎥ OCTAVE
$f1+f2$	700	⎥ PERFECT FOURTH ⎥
$2f1$	800	⎦ ⎦

The frequencies in this table again are the harmonics built on a fundamental of 100 Hz, but now there is a different set of harmonics present. And, the 700 Hz tone does not contribute to a musical interval; it adds a bit of dissonance. A person hearing this PERFECT FOURTH and then the PERFECT FIFTH of the last example might well identify the pitch of both as 100 Hz, but recognize the difference in timbre. Listen to the recording of these two musical intervals and hear this (Band 14).

• QUESTION 3.5.   Two frequencies are played, 150 Hz and 200 Hz. Reducing the ratio of these frequencies to their smallest integers gives: 150/200 = 3/4. Table 2.1 in Chapter 2 identifies this ratio of frequencies to be a PERFECT FOURTH. Do the analysis for the complex wave, the beats, and the combination frequencies and collect this information in a table.

Are the same named musical intervals heard as found in Example 2.)?

Part 1: Rules

What is the frequency of the fundamental of this harmonic series? •

Example 3.)   Two tones are played. One has a frequency of 500 Hz and the other 612 Hz. What do you hear?

First, note that these tones do not form a musical interval. The ratio of the frequencies is

$$500/612 = 125/153,$$

and this second ratio is both the ratio of the smallest possible integers for these tones, and is not one of the musical intervals.

A graph containing one cycle of the complex wave must include 125 cycles of the 500 Hz tone and 153 cycles of the 612 Hz tone, impossible to draw and understand. The complex wave's period is the time for 125 cycles of the 500 Hz tone and 153 cycles of the 612 Hz tone: 1/4 second. The complex wave's frequency is 4 Hz, a pitch too low to be heard. A harmonic series built on this fundamental would have the played tones as its 125th and 153rd harmonics.

The combination frequencies are

FORMULA	*COMBINATION FREQUENCY*
$2f1$	1224 Hz
$2f2$	1000 Hz
$f1 + f2$	1112 Hz
$f1 - f2$	112 Hz

There are no beats, but not for the same reasons as in the previous Examples. Remember that the period of a beat is the time it takes one of the played tones to make exactly one more cycle than the other. In this Example, that would be the time it takes the 500 Hz tone to complete N cycles and the 612 Hz tone to complete N +1. Working this out you will

Chapter 3

find that N = 4.41, and so both tones are in mid-cycle and not making the necessary start of a new cycle together.

• QUESTION 3.6. If you are algebraically inclined, put together and solve the equation for the value of N.

Hint: This equation in words is

(time for N cycles of the 500 Hz tone) = (time for N +1 cycles of the 612 Hz tone),

or, written another way,

(N periods of the 500 Hz tone) = (N +1 periods of the 612 Hz tone).

You take it from here. •

When all these results are put into our usual table of frequencies and intervals, and the mode number added to show how spread out the frequencies in this harmonic series are, then

TYPE	FREQUENCY, Hz (MODE)	MUSICAL INTERVAL
complex	4 (1)	
f_1-f_2	112 (28)	
Played, f_2	500 (125)	⎫
Played, f_1	612 (153)	⎬ OCTAVE ⎫
$2f_2$	1000 (250)	⎭ ⎬ OCTAVE
f_1+f_2	1112 (278)	⎪
$2f_1$	1224 (306)	⎭

Part 1: Rules

A meager, and not very interesting, pair of octave intervals is all there is. Listeners probably would not think this is beautiful music being played; they might even say it is dissonant, or even noise. Listen to recording of these tones (Band 15).

In each of these three Examples you are presented with two played tones. These could have come from two different instruments or might be from a single instrument simultaneously sounding two of its modes. For the former, the composer can decide upon the intervals and design the timbre; in the latter the listener must take what the laws of nature dictate.

DEMONSTRATION XII: Ghostly Combination Frequencies

Apparatus: Two Audio Oscillators, Two Loud Speakers

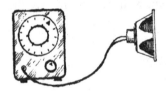 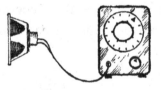

One of the audio oscillators is set to have its speaker play tone f1 at 500 Hz. The other plays tone f2 starting at 300 Hz and steadily decreasing the frequency to 100 Hz. You hear the f1 - f 2 combination frequency start at 200 Hz and increase to 400 Hz, and will be haunted by this ghostly increasing pitch caused by tones whose frequencies are not increasing. Listen to the recording of this (Band 16).

• QUESTION 3.7.   Sketch a graph of two waves having the same amplitudes on a pressure vs. time axes.

Chapter 3

One of the waves has twice the frequency of the other. On the graph identify a cycle of the complex wave.

If one of the given waves had a frequency of 100/s and the other had 200/s, what is the frequency of the complex wave?

Are there beats, and if so, what is the beat frequency?

Find the values of the combination frequencies.

Put all this information in a table such as you've seen, and identify the musical intervals present.

If one of the given frequencies were 100/s and the other were 201/s, how would the answers to the above change? •

3.8 Pitch-Frequency, Loudness-Sound Pressure, Timbre-Mode Content

This chapter ends with an additional discussion of the three pairs of terms: pitch-frequency will be examined again, loudness-sound pressure, and timbre-mode content. The first member of each pair is what a listener would use to describe what he or she perceives; the second describes the results of sound measurement with instruments.

Pitch and frequency are related; increasing one increases the other, but the relationship is not linear:

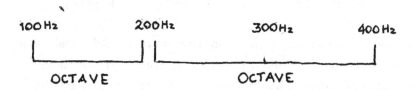

Part 1: Rules

Equal frequency intervals, in this case of 100 Hz, do not result in equal pitch intervals. Listen to this on the recording (Band 17).

Loudness and sound pressure are related in an even more complicated way. The ear-brain system is responsible for this. Loudness is what you hear, not what some instrument measures, although there are instruments, called sound level meters, whose response tries to mimic your ear-brain system. But because each of you has a different ear-brain system, the sound level meter readings are approximations to what you really hear.

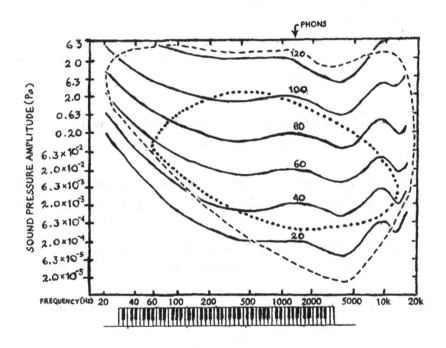

Figure 3.1 Sound Pressure Amplitude vs. Frequency

Loudness level curves (solid) compare the perceived equal-loudness levels expressed in units of phons to the sound pressure amplitudes. Approximate range of hearing is within the dashed boundary. Its upper edge is called The Threshold of Pain. Its lower edge is called The Threshold of Audibility. Approximate range of non-amplified music is within the dotted boundary. The scales are logarithmic to the base 10 in both the horizontal and vertical directions.

Figure 3.1 is an attempt to portray how the average person hears. The horizontal axis shows the frequency of a sound and the vertical one shows the pressure amplitude of the sound wave. The frequency scale is not linear. It is a logarithmic scale in which, as the piano keyboard shows, each octave spans the same distance on the horizontal axis. This is a device to show an equal amount of detail for each octave.

The frequency range is limited to audible sound (for people), and the pressure range goes from just audible to painfully loud sound.

The pressure is measured in metric units of Pascals (Pa). One Pascal is a small bit of pressure, and normal atmospheric pressure is about 1 x10^5 Pa. Note that the pressure amplitudes of sound waves, as shown in Figure 3.1, are much smaller than normal atmospheric pressure, and thus, sound waves in air do not change the meteorological atmospheric pressure much, hardly at all. Appendix A. "Sound Becomes Less Loud; a Closer Look at the Structure of a Gas" discusses and explains this in detail.

Figure 3.1 shows that sound pressure amplitudes increase by a factor of about a million from the least to the greatest loudnesses that people can hear in comfort. The decibel (dB) scale has been invented to express this spread within a smaller range of numbers. It, too, is a logarithmic scale, and you can see that each factor-of-ten pressure increase occupies the same interval of distance on the vertical axis.

The formula for converting pressure into dB is

$$dB = 20 \log_{10}\left(\frac{p}{20 \times 10^{-6} \text{ Pa}}\right),$$

where p is in units of Pascals.

Part 1: Rules

• QUESTION 3.8.   Convert the pressures shown in Figure 3.1 into dB and redraw this Figure with dB values for the vertical scale. Using the numbers on your redrawn Figure comment on the value of the phon relative to the dB. •

Timbre and mode content are probably the closest to being two descriptions of the same thing, but they are not identical. The timbre, or characteristic sound, is a perception and includes the combination frequencies produced by the ear-brain. Microphones will not find these frequencies in the air, and the complete perceived interval content is not measured by instrumentation.

CHAPTER 4.
MUSICAL SCALES AND TEMPERING

You now know the requirements for musical instruments: the frequencies of their modes must be integers times the frequency of their first mode. You also know that strings possess this property, so let's begin with strings. There are many stringed instruments and I'll divide them into two groups: those that play preset pitches and those on which the player can exactly choose the pitch. Both of these groups are important. The former includes the piano, harp, and all the fretted string instruments. The violin family dominates the latter; the player can make the string the length he or she wishes.

The world of preset instruments intrudes rudely into our ideal scheme of musical intervals. If we are to have pianos able to play in any key, the just tempered musical intervals (ratios-of-integers) must be abandoned.

4.1 Piano Design: Good-Bye, Just Tempering; Hello, Even Tempering

Your task is to design a simple piano. It need have only a four note scale: do, mi, sol, do'; but let's keep the just tempered frequency ratios. $f_{do} : f_{mi} : f_{sol} : f_{do'} = 4 : 5 : 6 : 8$. In Chapter 3 you found that $f_1 : f_2 = 1 : 2$ could also be written as fractions:

$$f_1/f_2 = 1/2.$$

The same notational method can be used here too. For example,

Part 1: Rules

$f_{do} : f_{mi} = 4 : 5$ can be written as:

$$f_{do}/f_{mi} = 4/5.$$

• QUESTION 4.1.   Show that the ratios of the frequencies for the just tempered notes do, mi, sol, do' are 4 : 5 : 6 : 8.

Hint: Table 2.1 is useful. In fact, it shows directly that the ratio of the frequencies of do to mi is 4 : 5. The other ratios are less obvious. •

And finally you must make your design so that a scale can begin on any note. Sounds easy?

Your piano's lowest pitch will be 100 Hz, and the keys for the first two octaves look like this:

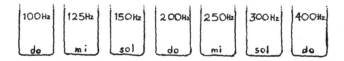

Playing a scale beginning on the second note requires these keys,

and a scale starting on 156.3 Hz, and so on, needs these:

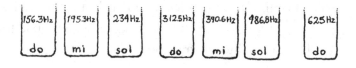

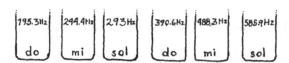

Chapter 4

Our original two octaves of 7 keys now must contain 16, and more will be needed as more scales begin. Your design task becomes truly impossible when a thirteen-note chromatic scale is specified. So many keys must be added that a player's hand can no longer span a PERFECT FIFTH. However, some clusters of similar frequencies have appeared above.

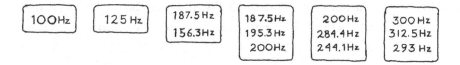

This suggests that you might try to put all the notes in a cluster into a single note whose frequency is some average of their values. This "fix" results in fewer keys, but destroys the just tempering's integer ratios of the musical intervals. For example, $f_{do} : f_{mi}$ will no longer always be 4 : 5. Is it worth it? Your ear must be the judge.

This process of adjusting the frequencies so that clusters can be replaced with a single note is a kind of tempering of the notes of the scale. There are many ways to temper them; you've already seen just tempering. Another could be based on simple averaging of the clusters' frequencies as suggested above, but it is not the accepted way to solve the problem of being able to play in any key, i.e., to begin a scale on any note.

The accepted tempering scheme is called **even tempering**, so named because each note in a scale is a constant pitch above the previous one. This automatically produces tones with which you can start a scale on any note, i.e., all the notes are the same pitch interval apart. They are not equal frequency intervals apart though. 100 Hz to 200 Hz and 200 Hz to 400 Hz are equal pitch intervals (an octave), but they are not equal frequency intervals. The new pitch intervals will be the semitones in the even tempered scale. These changes may seem to be asking for a lot, but you will find that the even tempered scales are not all that much different from the just tempered ones.

Part 1: Rules

Even tempering keeps the octave interval when the frequency is doubled. However, it adds the requirement that there is the same pitch interval between successive semitones; that is to say, the ratio of the frequencies between them is always the same. For example:

$$f_{di}/f_{do} = f_{re}/f_{di} = f_{ri}/f_{re} \ldots = \text{(a constant)}.$$

This ratio is a number; what is it? For the time being let's call this number, x, and work out a method to find its value. Now you know that

$$f_{di}/f_{do} = f_{re}/f_{di} = f_{ri}/f_{re} \ldots = x.$$

And, that this requires that

$$f_{re} = x f_{di} = x (x f_{do}) = x^2 f_{do}.$$

Here is the twelve semitone scale with the frequency of the lower do represented by the letter f, and the octave higher do' frequency by $2f$. This preserves the octave interval, being between two notes, one with twice the frequency of the other.

do	di	re	ri	mi	fa	fi	sol	si	la	li	ti	do'
f												2f
f	xf	x^2f	x^3f	x^4f	x^5f	x^6f	x^7f	x^8f	x^9f	$x^{10}f$	$x^{11}f$	$x^{12}f$

The frequency of each note is x times the frequency of the previous note. This keeps the semitone interval constant. Each note has the frequency of the previous one multiplied by x; and so after going from do to do', twelve intervals and twelve multiplications by x, you find that $2f$ is the same as $x^{12}f$. So, x^{12} equals 2, or x multiplied by itself twelve times is 2. This requires that x = 1.05946. Even tempering requires that you must increase the frequency by about 6%

Chapter 4

when going from one semitone to the next higher one. This is not the same number of cycles/sec for each semitone change; it is the same percentage change. The number of cycles/sec change increases as the pitch increases.

Here is a comparison of some notes' frequencies with just and even tempering.

NOTE	JUST TEMPER FREQUENCY	EVEN TEMPER FREQUENCY
do	100 Hz	100 Hz
mi	125 Hz	126.0 Hz
sol	150 Hz	149.8 Hz
do'	200 Hz	200 Hz

Even tempering looks pretty good. Play the recordings of these tones and intervals using these frequencies and hear that they sound all right, too (Band 18 and Band 23). J. S. Bach (1685-1750), an early proponent of even tempering, let the listener decide for him or her self, also. His *The Well-Tempered Clavier* contains pieces in every key signature to be played on an even tempered instrument.

Even tempering allows you to calculate the frequency of any note after you have specified the "first" note's frequency. Our presently accepted pitches are built on a first note's frequency of 440 Hz. Figure 2.2 shows the notes using even tempering and an A_4 frequency of 440 Hz. This is a rather recent specification; Europe and North America finally agreed on this value in 1939. Table 4.1 gives a partial chronology of this pitch until 1880.

Part 1: Rules

Frequency for a_4	Date	Particulars	Character
370.	—	Ideal lowest, or zero point	Lowest
374.2	1700	Lille, organ of L'Hospice Comptesse	church
376.6	1766	Paris, from model after Bedos	pitch
393.2	1713	Great organ in Strassburg cathedral	Low
395.2	1759	Organ at Trinity College, Cambridge, England	church
398.7	1854	Lille, restored organ of La Madeleine	pitch
402.9	1648	Paris, Mersenne's spinet	Low
409.	1783	Paris, court clavecins	chamber
414.4	1776	Breslau, clavichords	pitch
415.	1754	Dresden, Silbermann organ	
421.6	1780	Vienna, Stein pianos, used by Mozart	Mean
422.5	1751	England, Handel's tuning fork	pitch
423.2	1815	Band of Dresden opera, under von Weber	of
427.	1811	Paris, Grand Opera	Europe
427.8	1788	England, St. George's Chapel, Windsor	
433.	1820	London Philharmonic	Compromise
435.4	1859	Paris, Diapason normal of Conservatoire	pitch
440.2	1834	Scheibler's "Stuttgart Standard"	Modern
441.7	1690	Organ at Hampton Court Palace, England	orchestral
444.2	1880	United States "low organ pitch"	pitch
445.6	1879	London, Covent Garden Opera	and
448.4	1857	Berlin Opera	medium
451.7	1880	United States, Chickering's standard fork	church
458.	1880	United States, Steinway's pitch	pitch
474.1	1708	London, Chapel Royal, St. James	High church
484.2	1688	Hamburg, St. Jacobi Kirche, approved by Bach	pitch
503.7	1636	Paris, Mersenne's ton de chapelle	Highest
563.1	1636	Paris, Mersenne's chamber pitch	pitches
567.3	1619	North German church pitch	

Table 4.1 The Variations of the Pitch a_4

PART 2: MUSICAL INSTRUMENTS

CHAPTER 5.
STRINGED INSTRUMENTS--MAKING THE SOUND

5.1 Loudness, The Need for Soundboards and Soundboxes

You've seen that a string's modes of vibration are ideal for musical intervals, and thus, stringed instruments should be encouraged to play them. However, a string by itself doesn't contact much air, and so doesn't produce a loud sound. In order to be part of a practical instrument the string must drive a device that does vibrate lots of air: a soundboard or a soundbox.

The tuning fork is not a string, but can show that a soundboard increases loudness.

DEMONSTRATION XIII: Soundboard

Apparatus: Tuning Fork, Table Top

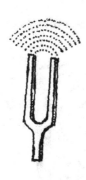
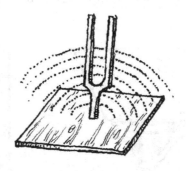

Part 2: Musical Instruments

The tuning fork by itself is not very loud. Its tines, although a lot bigger than strings, do not beat against much air. When the fork is touching a tabletop it causes the top to vibrate too and this big area is a much larger source of sound. The loudness increases a lot. Listen to this on the recording (Band 19).

The piano, viol and violin, guitar and banjo are examples of instruments with struck, bowed, or plucked strings that use this method. After you understand how these instruments make their sound you will also know the basics of musical bows, lyres, harps, lutes, citterns, zithers, dulcimers, clavichords, harpsichords, virginals, spinets, and ukeleles[4].

5.1.1 Soundboards and Soundboxes

Soundboards are two-dimensional surfaces. Their shapes and sizes give them resonant modes whose complicated vibrations must satisfy the conditions for a standing wave in two dimensions. This causes complex patterns of nodes and antinodes, as shown below, for a simple flat square plate.

DEMONSTRATION XIV: Modes, Nodes, and Antinodes, Chladni Figures

Apparatus: Audio Oscillator, Loudspeaker, Metal Sheet

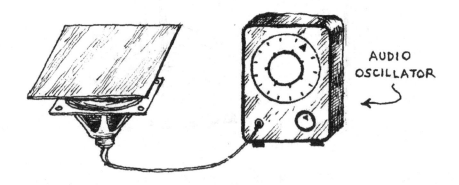

Chapter 5

A thin metal sheet above the speaker will play its modes when it is excited by the proper frequency from the speaker. Fine table salt scattered on the sheet will collect at the modes' nodes and be bounced away from its antinodes. The patterns formed are called Chladni figures. Ernst F. F. Chladni (1756-1827) demonstrated these by bowing the edge of the sheet at various locations.

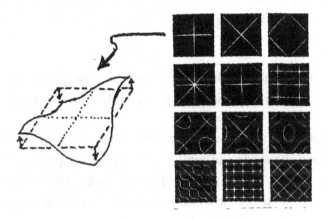

These are the Chladni patterns of nodes for a square sheet. Each one represents a resonant mode. The sketch shows a "strobe photo" of the sheet's motion during its first mode. The nodes are the dotted lines and the relaxed shape of the sheet is shown dashed.

Large soundboards will have low frequency 1st modes of vibration, and the frequencies of their higher modes will be closer together in higher scales (as you found out in Section 3.1). It is possible that several of these higher mode vibrations will be frequencies of notes the instrument is designed to play. If so, these notes will be extra loud. Or, these vibrations could have frequencies of notes that the instrument is not supposed to play. They might even play by themselves, independent of the musician's input. The former case will make the sound uneven, and the latter will change the instrument's timbre or even ruin the melody.

Soundboxes are three-dimensional volumes. Their sides are soundboards, but the vibrations of the enclosed elastic air are also sources of sound waves. Chapter 10 will discuss resonances in enclosed volumes of air that have one or more holes connecting them to the outside air. They are called Helmholtz resonators. Soundboxes contain soundboards and Helmholtz resonators, and the vibrations of both are present to help or hinder the sound. The combination of these vibrations is difficult to visualize, let alone to explain. Their frequencies, however, can be measured.

Wind instruments are almost entirely air enclosures, but their modes of vibration rate their own explanations, as you will see in Chapter 7.

In general, irregularly shaped soundboards and soundboxes are better. Their shapes and the struts and reinforcements that support and hold them together tend to prevent large vibrating areas. This raises the pitch of their 1st modes, which in turn puts fewer resonances in the instruments' scales. For example, if a soundboard acted like a harmonic oscillator whose 1st mode had a frequency of 50 Hz (which it probably wouldn't, but which will demonstrate the principle) the frequencies of its higher modes would be 50 Hz apart. If instead, the frequency of its 1st mode were 100 Hz, the higher modes would be 100 Hz apart; it would have half as many modes as the 50 Hz example and its 1st mode would be an octave higher.

5.2 Piano

Pressing a piano key throws a felt hammer against a string or strings that then vibrate with the frequency of the key's note. There are 88 keys, black and white, a semitone note apart, and they create the slightly more than seven chromatic scale octaves. Each bass note has only one string, the middle notes two, and the high three. The strings are steel, but not the same thickness nor tensions. If they were, each

Chapter 5

lower octave would need strings twice as long as those in the next higher one.

• QUESTION 5.1. Use the information and/or the equations in Section 3.1 to show that doubling the length of a string will lower its pitch by an octave. •

Part 2: Musical Instruments

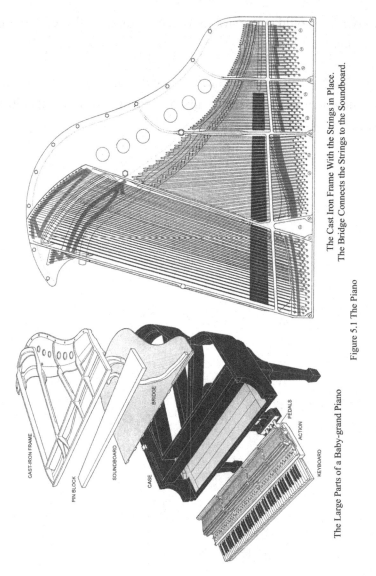

Figure 5.1 The Piano

The Large Parts of a Baby-grand Piano

Figure 5.1 shows the piano and names its parts. If its strings had identical thicknesses and tensions, the lowest pitch string would have to be about 256 times longer than the highest pitch one. The figure shows that this is not so. The low pitch strings are much too short. They achieve their low frequency vibrations by being wound with copper wire. This makes them sluggish and thus, they vibrate more slowly. Simply making the steel strings thicker will not work. They would become steel bars which when struck would "clink" with an even higher frequency.

Chapter 5

• QUESTION 5.2. Argue why it was claimed that the lowest pitch string would have to be 256 times longer than the highest pitch one.

Hint: $2^8 = 256$ •

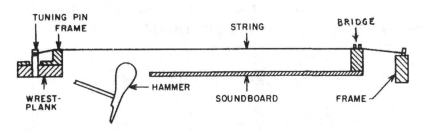

The above sketch is more or less a diagram of how the bridge connects the strings to the soundboard. It is not accurate because Figure 5.1 correctly shows the soundboard extending beyond the bridge. Two pegs in the bridge straddle each string and transmit its motion to the soundboard. The ends of each tightly stretched string are attached to the cast-iron frame, which supports them and the considerable total force of tension. This sub-assembly of frame and strings is dropped into the much less strong wooden case that contains the keyboard and the action that moves the hammers.

The piano's action throws the hammer when the key is pressed. The action is a clever system of levers and restraints that assure the hammer delivers a "hammer-like" blow to the strings. Rossing[5] shows and explains the action in all its mechanical glory.

The piano tuner tunes a piano by adjusting each string's tension, tightening to raise its pitch and loosening it to lower it. He or she does this by rotating the tuning pin; but cannot change the tension too much or the string will no longer work musically: too low a tension and the string will flop around rather than vibrate, and too much tension may stretch the string past its elastic limit. A piece of metal deformed too much will not spring back into its original shape. This "too much" is called its "elastic limit."

Part 2: Musical Instruments

5.3 Viols and Violins

The viol is the older of these two. The figure below shows the differences between them: six strings instead of four, fretted neck, sloping shoulders, and a flat back. The viol is held between, or on, the player's knees and its bow is held palm upward.

VIOL VIOLIN

Violins were developed from the viol. They come in four sizes: violin, viola, violoncello, and double bass, shown below from left to right.

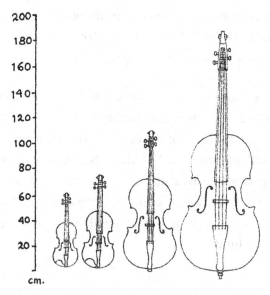

Chapter 5

The double bass is often made with the viol's sloping shoulders. It is bowed either palm upward or downward. Violins and violas are held under the player's chin. The violoncello player sits with it between his or her knees, and the double bass player stands or sits on a high stool behind it.

Viols and violins have similar shapes, and a description of the violin will suffice.

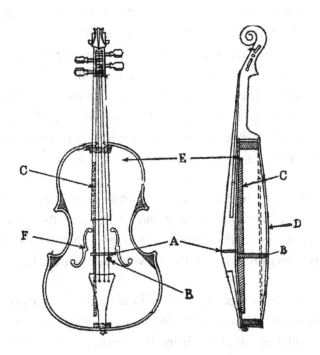

Part 2: Musical Instruments

The violin's strings are shortened when the musician presses them against the fingerboard. To show that this raises the pitch we return to an equation in Chapter 3, but use it a new way. Look again at equation (3.2):

$$f_n = v\, n/2\, L, \text{ with } n = 1, 2, 3, \ldots \tag{3.2}$$

In Chapter 3 this equation was for a fixed length, L, string. The frequency of mode n, f_n, was calculated. Violin strings are not fixed lengths. Length L shortens as they are pressed. However, equation (3.2) still applies. Now we look at what happens when n is kept constant and L changes. Equation (3.2) says that as L gets smaller, f_n increases. The pitch rises. The violin doesn't have frets and almost any shorter string length is possible. The requirement that its soundbox be able to vibrate at all these frequencies without any of them being one of its resonances is difficult or impossible to satisfy. It is a tribute to violinmakers and designers that this has been done as well as it has.

What makes the violin acoustically so interesting is its loudness. It is as loud as the piano without the piano's large soundboard, and much louder than the similarly shaped and sized guitar. The details of a violin's construction shown in the above figure reveal why.

First, note that the bridge, A, is full of holes and carved-away portions. The bow's sideways motion imparts many small horizontal plucks to the strings and they vibrate horizontally, almost parallel to the top plate, E, (or belly). The narrowness of the bridge in the left and right hand shaded areas between the cutaways permits the strings' motions to be easily transmitted into rocking forces at the bridge's feet. Next, observe that there is a sound post, B, and a bass bar, C, one under each of the bridge's feet. The sound post transmits the force to the back, D; and the bass bar couples the force on the other foot to the violin's belly. These rocking forces from the bridge's feet cause the air inside the violin to be compressed and "stretched," much like the air between the tines of the tuning fork. But, a much bigger volume of air is subjected to

Chapter 5

these distortions. The sound leaves through the f holes, F; and directly from the back and the belly. Hence the violin's loudness; and, indeed, without its soundpost, a violin sounds rather like a guitar[6].

Other methods have been tried to increase the violin's loudness and the directionality of its sound projection. The Horn-Violin was one attempt. The violin's soundbox has been replaced by a diaphragm and conical horns. The bridge is connected to the diaphragm, which vibrates into the horns. The smaller horn directs some of the sound directly into the musician's ear. The other parts on the left end include a chin rest and a chest rest.

5.4 Guitar and Banjo

GUITAR   BANJO

Part 2: Musical Instruments

These two instruments are about the same size, but have quite different timbres and loudnesses. The guitar has a soundbox whereas most banjos have a single soundboard. Acoustically, the guitar is a plucked and fretted violin without the sound post or bass bar. Fingers pluck both guitar and banjo strings which gives them much bigger amplitude vibrations than violin strings have. That the violin is louder is a tribute to its efficient soundbox and the excellent coupling the bridge provides between the strings and the soundbox. This was described in Section 5.3.

The guitar's soundbox contains two distinct vibrators: the air within the soundbox and the hole that connects it to the outside (a Helmholtz resonator), and the wooden top and bottom plates (soundboards).

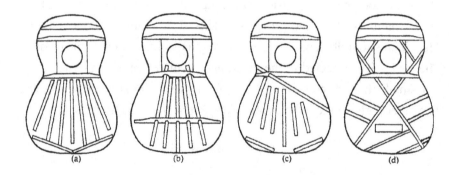

Here are some of the various designs for bracing guitar soundboard: (a) traditional fan bracing, (b) Bouchet (France), (c) Ramirez (Spain), (d) crossed bracing.

The top plate, as shown above, is braced with strips that strengthen it, but also lessen its flexibility and divide it into smaller areas. The guitar's bridge is stiff and low, and lies flat against the soundbox. Therefore, the bridge is less efficient at transferring the string vibrations to the already stiff soundboard. This correctly suggests that in addition to the soundboards, the Helmholtz resonator soundbox also determines the guitar's tone.

Chapter 5

The electric guitar does not have a soundbox, and thus, no soundbox modes. The motions of its strings are measured directly and electrically; and the signal is sent to amplifiers, modifiers, and finally to loud speakers.

The banjo's bridge is also lower than the violin's, but has two or three feet that stand on the soundboard. This bridge is more flexible than the guitar's, and does a much better job transferring the strings' motion to the banjo's single round flexible soundboard, which is usually made of thin plastic sheet. This relatively efficient string-bridge-soundboard coupling and the large area of the soundboard cause the banjo's loudness. Its "twangy" timbre is due to the round membrane's modes of vibration. The nodes of the modes of an ideal round membrane clamped at its edge are shown below. They are its theoretical Chladni figures.

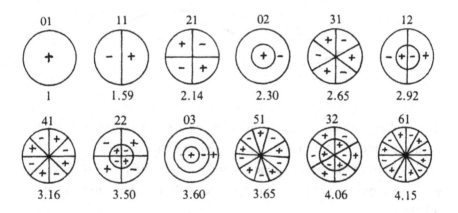

The plus and minus signs show which parts of the membrane are above and which are below its unstretched plane at the same time. A half period later, the plus signs become minus signs, and vice versa. The Chladni figures are lines of places on the membrane that never rise nor drop.

The digits above each figure tell how many straight and circular nodes are in that mode. For example, the upper left mode has one circular node (at its clamped edge) and no straight ones; thus, its designation as the 01 mode. Its motion is a simple up and down of the center of the membrane.

The 11 mode viewed edge on looks like this,

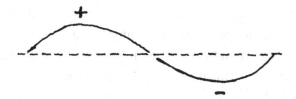

and a half cycle later looks like this:

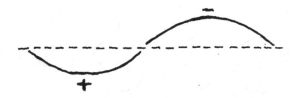

The rocking action of the banjo's bridge tends to drive straight-line modes 11 and 31. All the other modes have either up and down motion of the center of the bridge or have a node underneath the bridge. Neither of these types of motion is compatible with a rocking motion of the bridge.

The number below each mode is the multiple of that mode's frequency times the frequency of the 01 mode. Using the notation introduced in Section 3.1. of Chapter 3, we can write

$$f_{01} : f_{11} : f_{21} = 1 : 1.59 : 2.14.$$

For example, if the 01 mode's frequency is 100 Hz, the 11 mode's frequency will be 159 Hz. The 21 mode's frequency will be 214 Hz.

Chapter 5

The banjo's "twangy" timbre suggests that indeed its membrane's 11 and 31 modes are prominently present.

• QUESTION 5.3.   Show that a round membrane vibrating with its 11 and 31 modes will produce a MAJOR SIXTH musical interval. •

CHAPTER 6.
SOME MUSICAL PERCUSSION INSTRUMENTS

6.1 Membranes and Thick Strings

Perhaps the piano might be included here, but it isn't. It's too complicated; the hammers hit the strings, which then drive the soundboard. I want to limit this chapter to instruments whose soundboards are hit directly, and simplify everything into two groups: the thick strings and the membranes. The latter are mostly round Chladni plates: the drums, playing their modes. The frequencies of these modes are not integers times the frequency of the 1st mode; not even close; and unless special precautions are taken to allow only one mode, the sound will not meet the requirements for musical sound. Practical solutions include striking the drumhead at a particular place or touching it at the location of an unwanted mode's antinode.

Kettledrums have apparatus to change the tension of their membrane and an increase will raise the pitch. Some have individual tightening screws, or a single handle, or a pedal-operated device to do this. The handle and pedal adjustments evenly stretch or release the whole head (membrane). Uneven stretching, either on purpose or by accident, will cause nonsymmetrical Chladni-like figures on the head, which indicate

Part 2: Musical Instruments

another set of modal frequencies are present. They are too complicated to be discussed in this book.

The thick strings suffer from the same lack of integer relationships between their modes, but there are too many of them playing tunes to be dismissed as musical instruments. The xylophone, marimba, vibraphone, and glockenspiel all qualify in spite of their harmonic deficiencies.

A vibraphone is shown below. The bars are arranged like the black and white keys on a piano. The musician stands behind the "white-key" bars and beats them with mallets. Vibraphones have electric fans near the ends of the tubes. These fans rotate when the pedal is pressed and this causes the vibrato.

Chapter 7, Wind Instruments-Making the Sound, will explain why shorter tubes are necessary for higher pitches.

Marimbas and xylophones are similar to vibraphones, but without the fans. Also, the marimba has one more octave of bars than the xylophone.

Both the vibraphone and the glockenspiel have metal bars; the others use wooden ones.

A typical pattern of nodes and antinodes for the standing waves in thick string modes is

Chapter 6

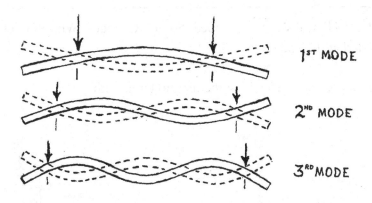

The frequencies of these modes are typically,

MODE	FREQUENCY
1	f
2	$2.7f$
3	$5.2f$
4	$8.4f$
...	...

Or using the notation introduced in Section 3.1 of Chapter 3.:

$$f_1 : f_2 : f_3 : f_4 : \ldots = 1 : 2.7 : 5.2 : 8.4 : \ldots$$

But notice from the above sketch that the locations of the outermost nodes are not in the same place for every mode. This suggests that if a thick string is supported at the nodes of its 1st mode, the higher modes will be clamped out. And this is how it is done.

Part 2: Musical Instruments

DEMONSTRATION XV. Tapped Sound Changes When Block is Supported at its Node

Apparatus: Wooden Block, Something to tap it with

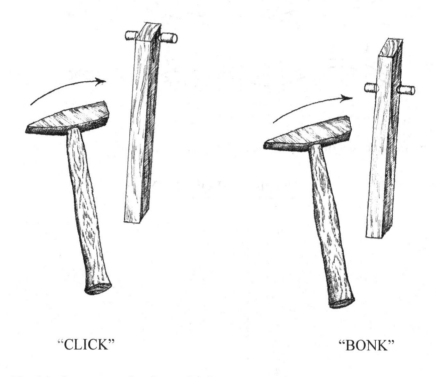

"CLICK" "BONK"

The block supported at its ends gives a much less musical sound when tapped. When the support is moved to the node of the 1st mode, tapping produces a real pitch. Listen to the recording of these taps (Band 20).

I dropped one of these blocks and that, too, made a musical "bonk." When it hit the floor it wasn't supported anywhere and the first mode sounded clearly. So, I gathered several blocks, all the same length, and then by trial and error, cut their thicknesses so that each hit the floor with a different pitch tone. Armed with this handful of instrument I let the blocks fall in a Fieldsian (W. C.) manner and played "Mary Had A Little Lamb". A hard floor is necessary.

Chapter 6

6.2 Music Directly from Mathematics

Some modern composers create a music outside my definition of musical sound. It is based on pure mathematics, the theory of combinations and permutations which tells the different ways you can put things together in groups, and perhaps, is the most elegant music of all. Here is an example.

Find three things you can hit, each producing a different sound. Call them 1, 2, and 3; and put a note on the staff closest to the pitch each makes.

Now, how many ways can you hit these three things, one at a time? Let's arrange them in groups of three for all possible permutations. Use the lowest numbers first. Here they are.

      1 2 3

      1 3 2

      2 1 3

      2 3 1

      3 1 2

      3 2 1

Part 2: Musical Instruments

Our music is practically composed. Let's make three bars of 6:8 time.

This is just one of the many schemes you could use to arrange these pitches. The next QUESTION asks you to arrange them using another.

The musical name for an ordered progression of tones is called a **change**. Sets of bells are popular instruments for performing changes.

• QUESTION 6.1.   Arrange the numbers in groups of three for all possible permutations. Use the highest numbers first. Show these permutations on a musical staff. How does your arrangement compare with the one shown above? •

CHAPTER 7.
WIND INSTRUMENTS--MAKING THE SOUND

Wind instruments are more difficult to analyze than the strings. There is no clear distinction between the vibrating object that sets the pitch and the device that produces the high and low pressure fluctuations in the air outside the instrument. Both of these are the column of air inside the instrument, and it's not easy to distinguish the proactive air inside from the passive air outside, which will be the medium through which the sound travels. Everything is air, but performing different functions. Let's begin with a general discussion of what modes can be present in the air inside the instrument. After this, you'll discover a way to sustain the vibrations of these modes, and finally, take a look at some particulars of real woodwinds and brasses.

7.1 Standing Waves in Air

Chapter 1 showed you deflections traveling through elastic media, and Chapter 3 introduced you to standing waves in strings. The traveling deflections going in opposite directions in the string combined to produce standing waves if the frequency of the deflections and the length of the string were just right. Those deflections were up and down, perpendicular to their direction of travel. This type of traveling wave is called a **transverse wave**.

The "deflections" in sound waves are pressure deviations caused by the air molecules bunching up and spreading out, and these are happening in the direction of travel. This type of traveling wave is called a **longitudinal wave**. The Slinky you saw in DEMONSTRATION V

Part 2: Musical Instruments

in Chapter 1 had this type of longitudinal motion: the compression happened in the same direction as it traveled through the Slinky. The motion of the coils' compression along the Slinky is a visible model of the motion of compression through air.

The Slinky is elastic (just what is needed) and the coils represent the air molecules. The bunched up coils represent a high, and the extended ones, a low. I will continue to use the Slinky model, but now add a driving to and fro oscillation to produce a traveling longitudinal wave. If, in addition, one end of the Slinky is fastened to a wall, a reflection will occur; if the driving frequency is right, a standing wave will result. DEMONSTRATION XVI will show this and then explain how to graph this motion. Finally, all this will be applied directly to sound waves inside wind instruments.

DEMONSTRATION XVI. Standing Longitudinal Waves

Apparatus: Slinky

DEMONSTRATION V in Chapter 1 used a piece of rubber tubing and then a Slinky to show how single pulses travel through elastic media. Let's again use the Slinky as a model for the invisible sound waves, but this time instead of sending a lone compression pulse down it, make a series of compressions and expansions by shaking it back and forth in the direction of its length at various frequencies. Each compression and expansion travels through the Slinky and reflects back from the fixed end, passing through the others. At some frequencies they do this in a way that makes it look like there are just local side-to-side vibrations. These are standing waves.

Chapter 7

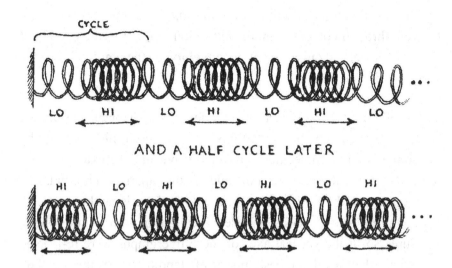

The small arrows show how the closely spaced coils will move to create other compressions and expansions. Identify a cycle of this motion and note that there are parts of the Slinky in between the compressions and expansions that don't move at all.

The sketch shows a half cycle of the motion. Each group of coils stays in the same place along the Slinky, and seems only to cycle between compression and expansion of its coils. The Slinky is resonating in one of its modes. Each compression and expansion is a cycling antinode, and each still place is a node.

Next are five graphs of the positions of the coils at five successive times. These graphs plot the coils' motion with their vertical axes representing the amount of the coils' distortion and the horizontal ones showing the distance along the total spring's length. The dashed curve shows the extreme amounts of compression and expansion at various distances; and the small arrows show the up and down directions of the graph as time progresses.

At some of these distances the coils compress and extend a lot and at others not at all. These are the locations of antinodes and nodes.

In between an antinode and its neighboring node, the coils continue to cycle through compression and extension, but not to the maximum values. The dashed curves form the **envelope** of the amplitudes of the various coils' motion.

The Slinky used this way is a good model of what happens in standing waves in air. The compressed Slinky coils represent highs caused by the compressed air molecules in a sound wave, and the extended coils represent lows caused by air molecules farther apart than normal. The locations along the horizontal axis, at which there is neither a stretch nor compression of the coils, represent locations of normal atmospheric pressure. So, if the vertical axes of these five graphs were re-labeled "pressure" they would be descriptions of standing waves in air. I'll do this in Section 7.2 of this chapter when I will be explaining standing waves in air, and not in Slinkys.

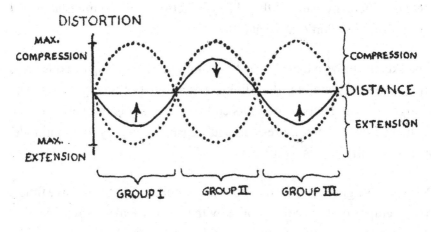

Graph 1.)

I have divided the distance into three Groups. Each Group contains the coils between successive nodes. The coils in Group I are now extended apart, but becoming less so. A short time before they had their maximum extension. Group II's coils are compressed together, but becoming less so. They previously had their maximum compression. Group III is

Chapter 7

doing the same thing as Group I. All the coils are moving toward their normal positions now.

Two neighboring Groups constitute a cycle, and the distance they occupy along the Slinky, or the graph, is this cycle's wavelength. Because the amplitude of the vibrations is also in the direction along the Slinky, this type of vibrational motion is more difficult to visualize. This and the following graphs separate the two different things happening in the same direction, plotting them on perpendicular axes, but you must keep in mind that everything is happening along the Slinky. This type of wave is a longitudinal wave. The waves you've seen in the rubber tubing, on the other hand, have their amplitudes up and down and not along the tubing. These are transverse waves. In either case, a graph of the displacement or distortion vs. distance will have the same look, and this look is a common feature of all kinds of wave motion. The sound wave's pressure vs. distance (or time) graphs you've seen have this appearance too, as they must because they are also members of the family of waves.

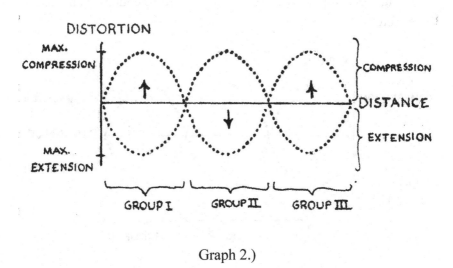

Graph 2.)

Graph 2.) happens a little less then 1/4 period after Graph 1.) . The Slinky has its normal shape, neither extended nor compressed, but only

Part 2: Musical Instruments

for this instant, and the graph is moving in the directions of the small arrows. Group I will begin to compress, Group II to expand, etc.

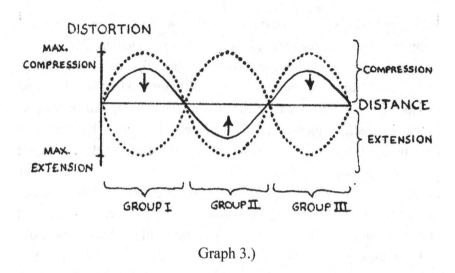

Graph 3.)

Graph 3.) happens 1/2 period later then Graph 1.). The coils in Group I have previously reached their maximum compression and are now becoming less compressed. Group II's coils are extended, but becoming less so, etc.

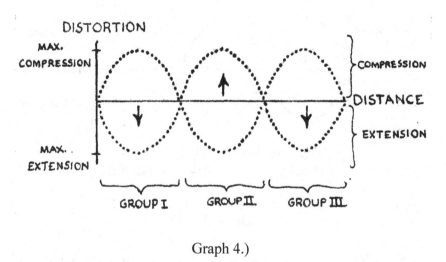

Graph 4.)

Chapter 7

Graph 4.) occurs 1/2 period after Graph 2.), and also a little less then 1/4 period after Graph 3.). The coils are again instantaneously in their normal positions, but will continue to extend or compress as the small arrows show.

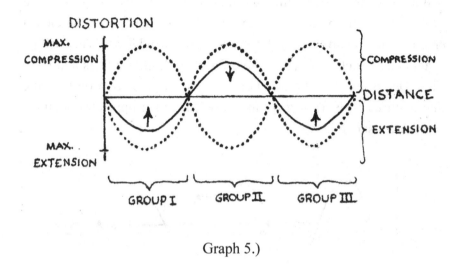

Graph 5.)

Graph 5.) is the same as Graph 1.); the coils have gone through a cycle of their motion.

These five Graphs describe the Slinky's motion but do not explain it. How can traveling waves produce standing waves? If two waves pass through each other their displacements combine and create the resultant wave. This is the same kind of interference you saw taking place to produce complex waves, but now the two waves have the same frequencies. A common example of this occurs when a wave interferes with its reflection, which assures that the second wave is the same kind, just traveling in the opposite direction. All common musical instruments use this method, either with transverse traveling waves in strings or longitudinal traveling waves in air, to create the resonant standing waves of the frequencies they play. See Chapter 10, "TWO OTHER KINDS OF WIND INSTRUMENTS," for some exceptions.

What causes a wave to reflect? If it is traveling down a string it comes to the fixed end and cannot go farther. The stretched part cannot stay stretched forever and transfers its stretch back along the string. This is also what happens when a rubber ball bounces off a wall. The ball stops while compressing and then springs back into shape while pushing away from the wall. The reflection of sound from the closed end of a tube is similar. A reflection at an open end is not. This will be examined in Section 7.4, "Pressure Pulses in Long Cavities and How They Determine the Modes." For now fix your attention on how the interference between two traveling waves moving in opposite directions can cause a standing wave.

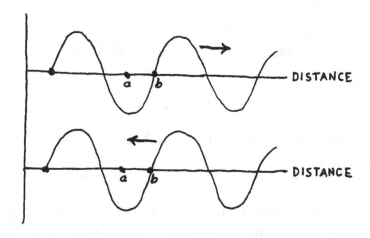

Here is a stopped action graph of two oppositely traveling waves. Although I've drawn them on two axes for clarity, they are two different waves traveling in opposite directions along the same horizontal axis. The arrows show their directions of travel. Whether they are in strings, in Slinkys, or in air makes no difference and the vertical axis is not labeled. The top wave travels to the right and the bottom one to the left, both with the same speed. At point a, both waves are having their lowest displacements. They interfere and add to produce a very low displacement there. A very short time later, the top wave has advanced a little to the right, and the bottom wave, an equal distance to the left.

Chapter 7

They are still both low at point a, but not so much, and the resultant wave is, therefore, less low, too. Some time later, the locations of the waves' zero displacements have reached point a, and the resultant wave displacement is zero. Still later, the high displacement parts of the waves cross at point a, and their addition produces a larger high. This continues and the resultant wave at point a cycles from low to high to low, back and forth. Point a is an antinode.

Look again at the top and bottom graphs. At point b both waves now have zero displacements. A little later, a low displacement part of the top wave has advanced to point b, but a high displacement part of the bottom one has also arrived there. The resultant wave continues to have zero displacement. This pattern of highs intercepting lows continues at point b, and the resultant remains zero. Point b is a node.

You should be able to determine that the nodes, or antinodes, are a half wavelength apart.

Graphs of standing waves are often, but not always, drawn with two curves, each one showing one of the extreme excursions of antinode motion. The horizontal axes of such graphs are distance, not time. All the graphs in Section 7.2 will use this format.

• QUESTION 7.1.   Pick some other point c on the distance axis and sketch a graph of the displacement vs. time of the resultant wave there. Does your sketch show that the graphs of displacement vs. time for any point you could choose will differ only in the value of the maximum displacement? •

• QUESTION 7.2.   Suppose everything is the same as above except the bottom wave's amplitude is about half the top's. This situation can occur if only part of a wave's amplitude is reflected.

Sketch the envelope of the resultant wave.

Part 2: Musical Instruments

Would there still be a standing wave? Why?

Are there any nodes in the resultant wave?

Are there any antinodes in the resultant wave?

How far apart are neighboring maxima? •

7.2 Boundary Conditions and the Modes of Ideal Cylindrical Cavities

This section explains that only certain frequencies can cause standing waves in the air inside cylindrical cavities, and tells you how to find the values of these frequencies. The frequency of any vibration that causes standing waves in any object (such as strings, Slinkys, cylinders, etc.) is called a **resonant frequency** of that object. All musical instruments are designed and made so that their playable notes are their resonant frequencies. This is usually accomplished by changing the vibrator's shape. Sections 7.3 and 7.4 will explain how standing waves are initiated and sustained in wind instruments.

Recall the Slinky-air model and liken the compressed coils to high pressure and the expanded ones to low. The still ones represent atmospheric pressure. The envelope of such behavior in air plotted on a pressure vs. distance graph for the two times when the antinodes' pressures are extremes would look like this.

Chapter 7

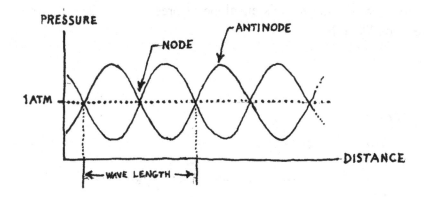

If these standing waves were happening in the air inside a cylindrical tube, open at one end and closed at the other, the open end would be clamped at atmospheric pressure, while there would be big pressure fluctuations at the closed end when the highs and lows hit there and bounced back.

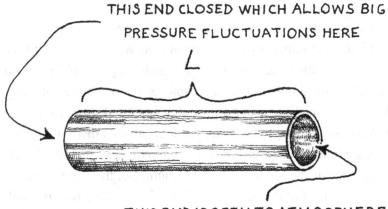

So, the closed end is the location of a pressure antinode and the open end is a pressure node. These are called the **boundary conditions** on the pressure and they must be obeyed in order to have a standing wave in this tube. The 1st mode's envelope of pressure vs. distance graph must look like this:

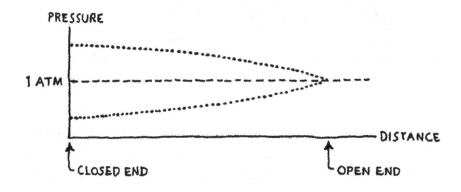

The previous pressure vs. distance graph showed several nodes and antinodes of a standing wave; this one shows only one of each. The closed end of the cylinder is the location of a pressure antinode and the open end is a pressure node clamped at atmospheric pressure because it is connected to the almost infinite atmosphere. These are the boundary conditions for a standing wave in a cylinder closed at one end and the graph shows the longest wavelength that satisfies them. Only one quarter of a whole cycle of the wave is necessary, and thus, the length L is one quarter of the 1st mode's wavelength. So, the wavelength of the 1st mode is $4L$. For the rest of Section 7.2 only the envelopes will be shown. This is a common way to indicate all the pressure variations that occur at a particular distance on a pressure vs. distance graph, and tries to introduce time as a variable in these graphs. Edward R. Tufte[7] would approve.

Chapter 7

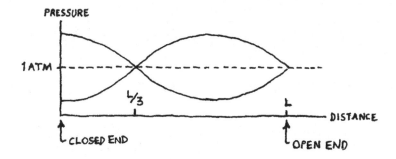

The next mode, the 2nd, which would obey the boundary conditions, has a shorter wavelength as shown in the above graph. This mode must also satisfy the cylinder's boundary conditions. The cylinder's length is still L, but now the graph above shows that a shorter wavelength is necessary and satisfactory. L is now 3/4 the length of this new wavelength, and this new wavelength is thus $(4/3)\,L$, quite a bit shorter than that of the 1st mode. The 2nd mode's frequency is thus higher. Note, also, that the atmospheric pressure node is 1/3 the way down the cylinder from the closed end.

And, the 3rd mode would look like this:

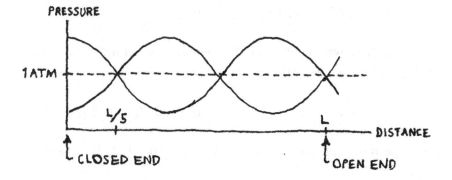

The next shortest wavelength that satisfies the boundary conditions for a standing wave in length L is shown. This wavelength is $(4/5)\,L$.

Put this information into a table.

Part 2: Musical Instruments

MODE	WL
1	$4L$
2	$\frac{4}{3}L$
3	$\frac{4}{5}L$

This table shows that the possible modes' wavelengths are given by the formula,

$$WL_n = 4L/n \; ; \qquad n = 1, 3, 5, \ldots$$

The possible frequencies are calculated from the general formula (2.3),

$$v = (WL_n)f_n,$$

and so,

$$f_n = v/WL_n = ((345\text{m/s})/4L)\,n, \qquad n = 1, 3, 5, \ldots \qquad (7.1)$$

Note that these wavelengths and frequencies are not the same as for a string, which you saw earlier as,

$$WL_n = 2L/n, \qquad n = 1, 2, 3, \ldots \text{ and}$$

$$f_n = ((\text{speed of wave along string})/2L)\,n; \qquad n = 1, 2, 3, \ldots \quad (3.2)$$

The different speeds of the waves through the air and along the string account for some of the differences in equations (7.1) and (3.2) for these frequencies, but it is the different boundary conditions that are responsible for the 4 and the 2, and that the integers, n, are odd for the closed end tube, and both odd and even for the string. The string is

Chapter 7

fixed at both ends and thus boundary conditions require a node at both ends.

Moving from the analysis of simple pipes to real instruments adds considerable complexity. The clarinet is more than a cylindrical tube closed at one end. It has tone holes; and as the picture in Chapter 9 shows, they seem to be grouped together near the center of the tube. Why not space them along the whole length? Figure 2.2 in Chapter 2 shows the range of a B ♭ clarinet: about 3 octaves. This requires that the clarinet play notes in several of its modes. The notes played using the 1st mode are simple enough to analyze. You just open up the tone holes and play the 1st mode of a successively shorter instrument. The notes played in the 2nd mode could use a register key located one third of the way from the reed. But, the location of the open end changes as the tone holes are opened. And therefore, so does the location of the needed register key. Where can you put it? QUESTION 9.1 in Chapter 9 requires you to examine, and to resolve, this dilemma. Let's continue now by comparing the modes in fixed-length cavities to those in strings.

If the cylinder (soon to be identified as a flute) is open at both ends, the boundary conditions are atmospheric pressure at both ends, i.e. pressure nodes there. The pressure vs. distance graph for the 1st mode of this tube is

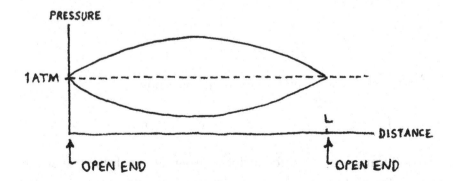

Length L is 1/2 of the longest possible wavelength that obeys the boundary conditions for our cylinder open at both ends. So, the wavelength of the 1st mode of this cylinder is $2L$.

And the graph for its 2nd mode is

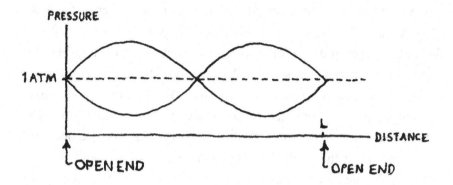

The wavelength of the 2nd mode is L.

And so on. These findings are tabled:

MODE	WL
1	$2L$
2	L
3	$\frac{2}{3}L$

and the formula for the modes' wavelengths is now,

$$WL_n = 2L/n, \qquad n = 1, 2, 3, \ldots$$

This is formula (3.1) we found for a string, although the waves are quite different. One travels back and forth down the string; the other

Chapter 7

back and forth in the column of air in the cylinder. The frequencies here are

$$f_n = ((345 \text{ m/s}) / 2L) \, n \, , \qquad n = 1, 2, 3, \ldots \qquad (7.2)$$

Again, this is the same form of formula as found for a string's modes' frequencies except for the different wave speeds.

DEMONSTRATION XVII: Modes of a Tube Open at Both Ends

Apparatus: Flexible Plastic Tube (Lasso d'Amore, according to P. D. Q. Bach)

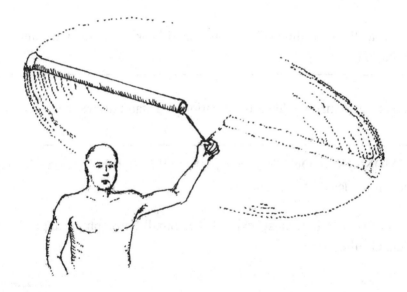

The tube, with a length, L, about 1 1/2 m long and with a 4-5 cm bore, is swung in a circle at various speeds. It plays tones, the resonances of the tube. The equation (7.2) predicts the frequencies of these resonances. I've called the first frequency f, and so equation (7.2) indicates that the higher modes' frequencies are $2f$, $3f$, $4f$, etc. Here they are, tabled, along with the musical intervals they make.

Part 2: Musical Instruments

MODE	FREQUENCY HEARD	FREQUENCY FROM THEORY		
1	not played	$\dfrac{345 \text{ m/s}}{2L} \cdot 1$		
2	$2f$	$\dfrac{345 \text{ m/s}}{2L} \cdot 2$	⎤ PERFECT FIFTH	⎤
3	$3f$	$\dfrac{345 \text{ m/s}}{2L} \cdot 3$	⎦ ⎤ PERFECT FOURTH	OCTAVE
4	$4f$	$\dfrac{345 \text{ m/s}}{2L} \cdot 4$	⎦	⎦

Listen to the recording of the tones, and hear the predicted intervals (Band 21).

You can also make a direct observation of a standing wave in a tube.

DEMONSTRATION XVIII: Microphone Probes Tube to Locate Nodes and Antinodes of its Resonances.

Audio Oscillator, Loudspeaker, Tube, Small Microphone taped to a stick. Oscilloscope

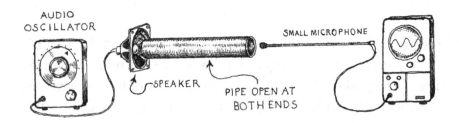

Adjust the speaker's pitch and loudness until it is playing one of the tube's resonant frequencies. As you change the frequency you can tell

when this happens; all of a sudden the sound near the tube gets louder. Insert the mike into the tube and note how far in it is when it is at a node or antinode. You can observe the displacement of the standing wave by looking at the height of the scope's display: lots of height at an antinode and practically none at a node.

The locations of the nodes and antinodes are as predicted.

Wind instruments are cylindrical, but there are also some with conical cavities. Both of these are common configurations. However, the explanation for the frequencies of a cone's modes defies the above analysis. You must wait for it until Section 7.4 where you will find out how the modes are sustained.

7.3 Sustaining the Tone

All wind instruments must have an open end; this is where the sound comes out. At this end, part of the pressure fluctuations are transmitted out into the room's air and the rest are reflected back down the cavity to keep the standing wave going. If there is only a small amount of reflection, the fluctuations in the cavity will decrease, perhaps quite rapidly. The instrument will be loud, but will quickly quit playing unless there is some way to sustain the tone.

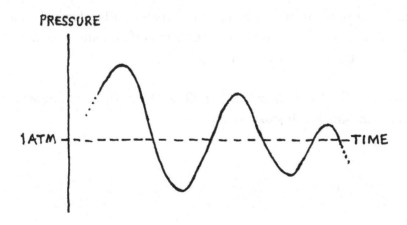

The above graph shows how the pressure at an antinode would die away. Note that the horizontal axis is time.

If the tone is to be sustained, pressure must be added at the same rate that it is lost. How is this done? First, let's find out when the pressure puffs must be added. Because the pressure is decreasing you might want to add puffs when the pressure fluctuations are at their lowest. This won't work. It will cause the fluctuations to dampen even faster, just what you don't want. The next graph shows why.

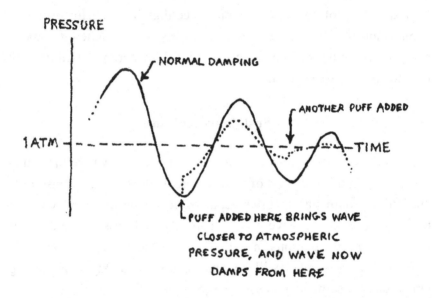

The dotted-line part of the graph, which has the puffs added when the pressure is lowest, dies away even faster than if no puffs were added. This is not the way to sustain the tone.

If, on the other hand, the pressure puff is added when the pressure is highest, the graph will look like this:

Chapter 7

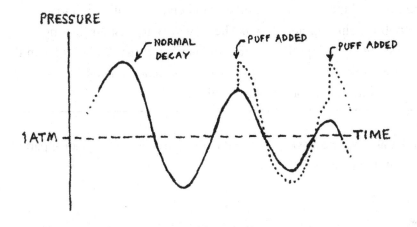

The dotted line is a strange shape, but it is sustained. Sustaining oscillations by adding energy this way is common to vibrating systems. Here is a demonstration of this being done to a mechanical oscillator.

DEMONSTRATION XIX: Sustaining an Oscillation

Apparatus: Spring and Bob, Rubber Band Chain

Part 2: Musical Instruments

Lightly hold the end of the rubber band chain and give it a tug to start the bob moving up and down. The best way to sustain this motion is to give the chain a pull when the bob is uppermost. You pull when the chain is pulling hardest on you. This is an exact analogy to adding a pressure puff when the pressure is highest.

Now that you know when to add the puff, you must design an apparatus to do it. Consider this:

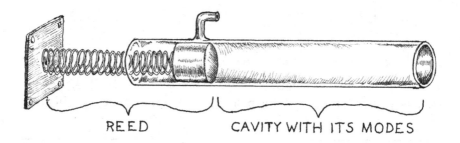
REED   CAVITY WITH ITS MODES

A close fitting but freely sliding piston is attached to a spring as shown. The tubular cavity is attached to a source of compressed air through the nozzle. The cavity, closed at the piston end, plays its modes, and high and low pressure fluctuations occur at this closed end. When the pressure there is high, the piston will be driven to the left and a puff of compressed air enters the cavity. When the pressure is low, the spring pushes the piston to the right, shutting off the air supply. The puff is added when the pressure there is highest, just what you want. The oscillations in the cavity are sustained even though some sound leaves through its open end. If more sound broadcasts out, a bigger air puff must be added. If you are the player (source of compressed air), you must blow harder.

This device is called a **cavity-controlled oscillator**. The oscillator in this case is the reed. Of course, your job as the player is more than just being air supplier. You must also start the fluctuations in the cavity and adjust its size so that the mode is the pitch you want.

Chapter 7

The above sketch shows the cavity as a cylinder open at one end. It could also be a cone or any cavity having resonant modes produced by standing waves. The new additions are the reed and the compressed air supply.

7.4 Pressure Pulses in Long Cavities and How They Determine the Modes

The explanations of the modes and their frequencies in the Section 7.2 give the right answers, but they leave important questions unanswered. How does the sound wave reflect at the open end of a cavity? How can the sound wave, with its high and low pressure parts, leave the cavity at its open end if the pressure there remains a constant atmospheric pressure? You have just read about the action of the reed that lets puffs of high pressure into the cavity. Where do the low pressure regions come from?

It's time to take a more realistic look at what's going on inside a cavity. You will find that the high pressure leaving the open end creates a low pressure pulse just inside and traveling back toward the reed. It is the combination of the leaving high and the forming low that keeps the pressure atmospheric at the open end. The momentarily open reed creates the highs in the cavity; the lows are made at the open end. You will also see that lows leaving the open end of the cavity create reflected highs there.

Now the pressure vs. time graphs will indicate what's happening at the reed end of the cavity. And, now, the pressure highs and lows will be represented by discrete pulses instead of being, or being based on, the smoothly changing curves you've previously seen. The exact shape of the pressure vs. time graphs for the sound waves inside the wind instruments depends on how long, and in what manner the reed opens and closes. It will be different for different reeds, and probably have some shape between the extremes of pulses and smoothly changing

Part 2: Musical Instruments

curves. I'll show some examples with both types. You will see that applying either type produces the same general conclusion; and therefore, the real situation, if it's somewhere between these types, will have this conclusion also. This is called **continuity**, a useful scientific argument if correctly applied.

7.4.1 Cylindrical Cavity with a Reed at One End and the Other End Closed

Let's begin by following one cycle of the motion of a puff in a cylindrical cavity having a trapdoor reed at one end and being closed at the other. This is a cavity closed at both ends and not a musical instrument. There is no way for the sound to get out; and also there will not be any low pressure pulses because there is no open end. Nevertheless, this is a good starting place for the analysis of more realistic woodwinds. Here is a cross sectional side view of this instrument, and below it the route of the high pressure puff during a cycle.

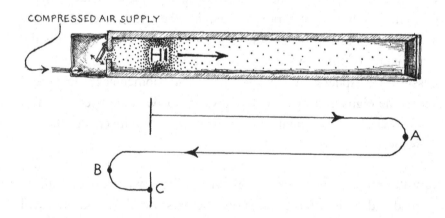

The puff has been injected through the now closing trap door, the reed, and is on its way down toward the closed end, point A on the pathway. At point A the puff compresses even more, and with no place to go bounces back down the cylinder toward the reed. When it reaches point B it pushes the trap door open and the next puff enters. This increases

Chapter 7

the high pressure region there that again expands back down the cavity to the right. At point C the puff has completed one cycle of its 1st mode. It is again a high pressure region at the same place in the cylinder, and traveling in the same direction as it was when we started examining it. The period is the time it takes the puff to make this journey. The frequency of the wave equals the frequency of the opening of the reed.

If for some reason the trap door opens before or after the high arrives there, the puff will not reinforce and sustain the previous one. In that case the confusion of puffs returning to and coming from the reed will not have the synchronization needed for a constant frequency and musical sound will cease. However, as long as it is the returning high that opens the trap door, this will never happen and the puffs will be added when the pressure at the reed is highest, just what is needed to sustain the traveling high pressure region. The reed, cylinder, and air source have become a cavity-controlled oscillator.

Here is a pressure vs. time graph of what's happening at the trap door end of the cylinder.

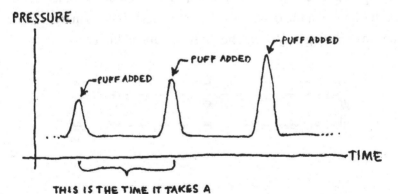

This is not a very interesting graph, and it shows each successive return with greater pressure. The incoming puffs just keep adding to what is already there.

You've seen that the frequency of the wave is also the frequency of the opening of the reed. You've also been shown that the reed opens whenever a high pressure region hits it. So, the frequency of the wave is the frequency with which highs arrive at the reed. The highs will return sooner in a shorter cylinder and, therefore, its length determines the wave's frequency. This conclusion was made differently in the Section 7.2 where its argument was based on the necessity that the boundary conditions be obeyed. It is reassuring to see that these two methods of analysis give the same conclusions.

7.4.2 Cylindrical Cavity with a Reed at One End and the Other End Open

Now you will see how an open end of a cavity can cause the needed reflections, and how the pulse reflects by changing high pressure pulses into lows and vice versa. Here is a cross sectional view of a cylindrical cavity open at one end and with a trap door reed at the other. I won't show the source of the compressed air any more, but it's there. The path of the wave during a cycle is shown below the cylinder.

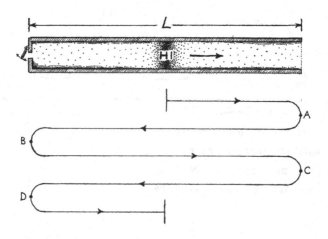

Chapter 7

The high-pressure pulse has been reinforced by a puff from the reed and is moving to the right down the cavity. It is constrained by the cavity walls until it arrives at the open end at point A on the path sketch. There it is no longer enclosed and expands (with the speed of sound) into the open air. Because the air has moved outward, it leaves a low pressure region and the neighboring high pressure region rushes in to fill it. Part of this comes from the cylinder, and this deficiency is in turn filled from the next layer of air in the cylinder. The low pressure deficiency moves back along the cavity toward the reed. And so, a returning low pressure region is created at the cavity's open end. When this low gets to the reed, at point B, it compresses a bit but not enough to open the reed. It bounces off the closed reed and heads back along the cylinder toward the open end, which is also point C on the path sketch. At point C it fills, even over fills, from the open air and this high pressure region travels back through the cylinder, to point D, where it pushes open the reed, and with the added puff there, regains its original high pressure. This high moves down the cylinder to the right and completes the cycle for a cylinder open at one end and with a reed at the other. Note that at A, a high is expelled from the cavity, and at point C, a low is expelled. These pressure fluctuations are the sound you hear.

This cycle takes two round trips through the cylinder. This is shown below on a one-dimensional graph of these events. The horizontal axis is the distance along the cylinder of a high pressure pulse as it moves through the cavity, changing into a low and again becoming a high. We will pick up the motion of the high puff halfway down the cylinder as shown in the above sketch.

Part 2: Musical Instruments

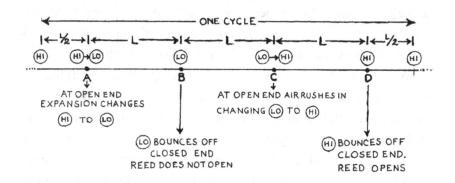

So, the original high spends half the cycle being a low. This is just what you've seen in earlier graphical representations of simple sound waves whose pressure changes from high to low and back during a cycle.

Note that one cycle of this 1st mode occupies $4L$, i.e.,

$$WL_1 = 4L,$$

which is the same result gotten in Section 7.2 from applying the boundary conditions. The puff is added once during this cycle, with the wave's frequency, as it must be. A cycle in this cylinder takes twice as long as in a cylinder closed at both ends. The reed must wait for two round trips between openings, and the frequency of the 1st mode is an octave lower than for the completely closed cylinder's 1st mode. Happily, all this agrees with the results of the analysis based on satisfying boundary conditions, and you should be convinced that they are equivalent arguments.

A pressure vs. time graph for what's happening at the reed during the 1st mode of this cylinder has both highs and lows. The peaks and troughs are drawn narrower than they really are in order to make the graph, and the succeeding ones, easier to read.

Chapter 7

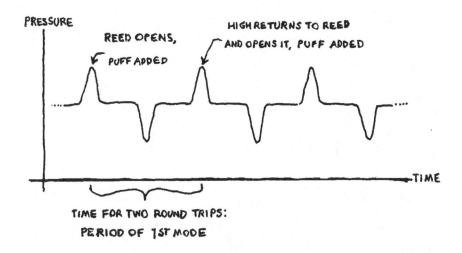

Can this apparatus sustain several frequencies at the same time? Yes, if the other frequencies are integer multiples of the 1st mode's frequency. In such a case the reed opening with the 1st mode's frequency would also give some additional high pressure to every other high of the 2nd mode, and every third high of the 3rd mode, etc. All of these modes are producing sound outside the cavity, and must have their highs replenished, but only the 1st mode will have its high refilled in every one of its cycles. The 2nd mode must wait two of its cycles for the puff it needs, and the 3rd mode must wait three of its cycles for its puff. This tends to make the higher modes have smaller pressures amplitudes, and thus the instrument's sound will have less loud higher modes.

Some modes, even though they meet the integer multiple frequency requirement, will not be sustained, however weakly, in a cylindrical cavity open at one end. Here is an example.

The graph below shows time variations of the pressure at the reed of 1st and 2nd modes, which have a frequency ratio 1 : 2. The ratio of the periods is 2 : 1. Look again at equation (2.1) if you need a reminder of the relationship of frequency to period. I've shown part of this graph in the smoothly changing curve representation. Its complex curve shows the same pattern of highs and lows as the pulses.

Part 2: Musical Instruments

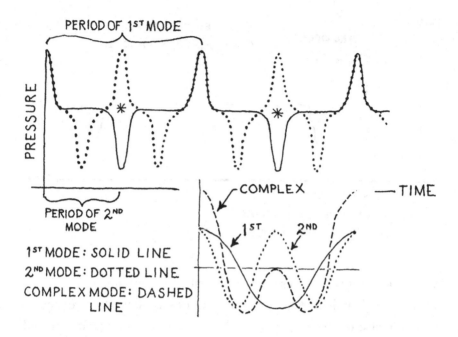

Well! Although they start together, by the times shown with "*" the higher frequency oscillation needs a puff, (it is high there), but at that same time, the 1st mode has low pressure. These two pressures add (by superposition) to atmospheric pressure and the reed will not open. There will not be enough pressure then to overcome the pressure in the air source that keeps the reed closed most of the time. The graph shows that without this puff added, the 2nd mode is faced with the impossibility of producing two lows in a row at the reed, and will not continue. It will quickly die away, and an open-ended cylinder will not support a frequency twice that of the 1st mode's. The 2nd mode, if it exists, must have a higher frequency. You have seen that applying boundary conditions to this cylinder produces equation (7.2), which specifies that the ratios of the modes' frequencies are 1 : 3 : 5 : ; so, let's try this and draw the pressure vs. time graph for the first two modes when 2nd mode's frequency is three times the 1st mode's.

Chapter 7

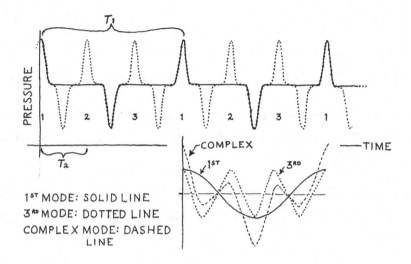

The 1st mode's pressure vs. time graph is indicated by the solid line. The three high pulses that occur during one period of the 1st mode are shown with a dashed line and numbered 1, 2, and 3. The highs and lows do not interfere with each other. Each high can open the reed, and both modes are sustained. The 2nd mode's frequency is three times the 1st mode's. This is new, and is unlike the string whose 2nd mode's frequency was twice the 1st mode's. The 1st harmonic of the sound from an open-ended cylinder has a frequency three times its fundamental. This agrees with equation (7.1), which was derived by applying boundary conditions to the wave. The complex curve shows that the intermediate high pressure pulses are less high. This suggests that some of the puffs added for the 3rd mode are smaller, and that the 3rd mode will be less loud.

Continuing these graphs would show that a frequency four times the 1st mode's would not be sustained and that the 3rd mode's frequency is five times the fundamental's. Again the modes' frequencies agree with those gotten from applying boundary conditions. Thus the resonant frequencies of a cylinder open at one end are in the ratios 1 : 3 : 5 : 7 ... If the fundamental frequency is 100 Hz, the other modes for this cavity will have frequencies 300 Hz, 500 Hz, 700 Hz, etc. The musical

Part 2: Musical Instruments

interval between the fundamental and the next mode is an octave and a fifth.

•QUESTION 7.3. Sketch the graphs for pressure vs. time at the reed end of an open-end cylinder for the following cases.

Two modes that start together, but one's frequency is four times the other's. Argue that your graph shows that the higher frequency mode is not possible.

Two modes that start together, but one's frequency is five times the other's. Argue from your graph that the higher frequency mode will be allowed. •

7.4.3 Conical Cavity and Cylindrical Cavity Open at Both Ends

Some woodwinds play as if they were cavities open at both ends. These include those with conical cavities and an instrument that is open at both ends, the flute. Flutes are cylinders with both ends open: the blown embouchure is one, and the openings at the other end are the other.

Here is a sketch of a conical cavity and its trap door, and the puff's route during a cycle.

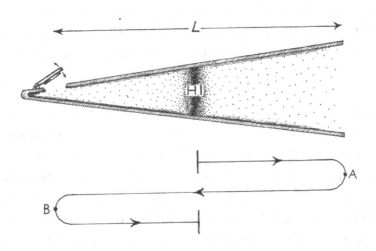

Chapter 7

The conical instruments have the reflection at the open end, A, which changes a high into a reflected low, but the diminishing cross section squeezes the low into a high again by the time it gets back to the reed. This opens the reed and lets in a puff to sustain the sound, and the length of a cycle is $2L$, twice the length of the cavity. This is the same result, although certainly not the same pressure mechanisms, as found in Section 7.2 for cylinders open at both ends. So, equation (7.2) also specifies the ratio of the modes' frequencies for conical cavities: 1 : 2 : 3 : 4.... Because a cycle here contains one round trip through the cavity instead of the two needed by a cylinder closed at one end, the wavelength of the first mode of a conical cavity is half as long as for that of a closed end cavity the same length. The fundamental frequency for this cone will be an octave higher than for the cylinder.

You may be wondering what is the difference between a cone and that flared tube called a bell at the end of brass instruments. The difference is what's happening at the open ends. At the end of a cone, the cross sectional area changes abruptly from whatever the area of the cone is, to the infinite area of the open air. The bell's cross-section is shaped to make a more gradual transition there. This reduces the amount of reflected wave at the open end, and more sound escapes. Brass are louder, but the interplay between the reed (the lips) and the waves in the cavity is lessened. This gives the player more control of the sound and allows him or her to make slurs between notes, which are quite a bit more difficult to do with woodwinds. Section 8.2.1, "The Bell," in Chapter 8 describes this effect of the bell and another important one.

The trap door is always open in flutes, and when the reflected low reaches it, a puff of the player's higher pressure air enters. The low becomes a high. The pattern and timing of the highs and lows is the same as for the cones, and so are the ratios of the frequencies of the modes. Here is a sketch and a one-dimensional graph of this process.

Part 2: Musical Instruments

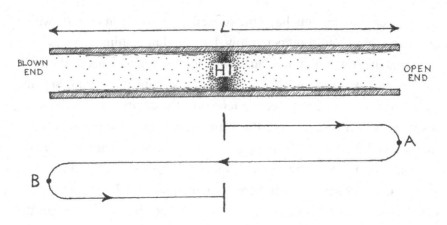

A puff has been added at the blown end and the high is moving down the cylinder. It will be partially transmitted-partially reflected at the open end A, as usual, and return down the cylinder as a low. When the low reaches the blown end at B, another puff will be added from the higher pressure blown air. So, a pressure change occurs at both ends, from high-to-low at A and from low-to-high at B. The one-dimensional graph of a cycle of these flip-flops is shown below.

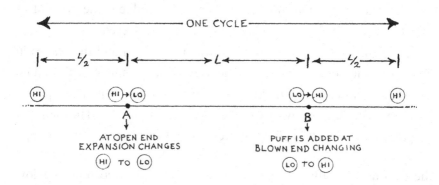

The cycle's length is $2L$, just like the cone's. The ratios of the modes' frequencies are: $1 : 2 : 3 : \ldots$; and the modes' frequencies are also specified by equation (7.2).

Chapter 7

DEMONSTRATION XX: Construct a Cylinder with a Reed, and Play It

Apparatus: Sheet of Paper, Scissors, Plastic Tape

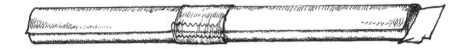

Construct this simple instrument by wrapping some thin paper around a large pencil and taping it to form a cylinder. Cut away a part of one end of the tube leaving a flap attached by a bit of its paper. Put the open end in your mouth and gently inhale. The flap (reed) vibrates open and shut, and the instrument plays a raucous tone. Start with a long cylinder and then shorten it. The changed pitch shows that the cavity is controlling the reed.

• QUESTION 7.4   Calculate the frequencies of the first three modes for the following shapes, each 1.5m long:

   - a cylinder open at one end and a reed at the other

   - a cylinder open at one end and blown at the other open end

   - a cone open at the big end and a reed at the other

Name some instruments with each of these characteristics. Exclude brasses that are made out of combinations of cylindrical and conical tubing. •

7.5 Real Reeds

You have been introduced to "trap door" reeds that are opened by pressure pulses in the instrument. Real reeds include this type and others, as well. Figure 7.1 shows a classification of them for most of the wind instruments.

Part 2: Musical Instruments

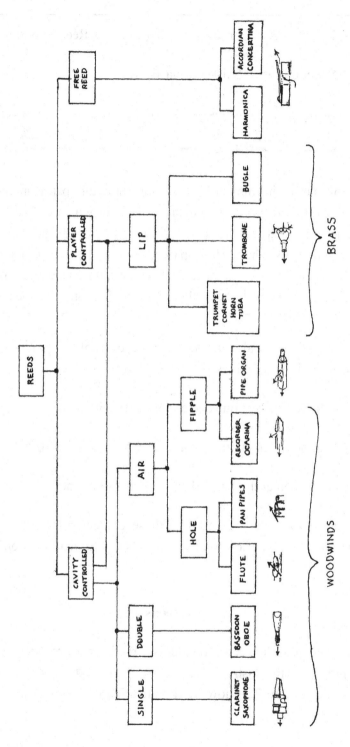

Figure 7.1 Some Wind Instruments According to Their Reeds

Chapter 7

Real reeds may or may not faithfully follow the orders of the cavity's pressure fluctuations. This depends on the inertia of the reed and the size of its area exposed to these fluctuations. Massive reeds will not quickly change their vibration frequencies; but if they have large areas in contact with the pressure fluctuations in the cavity, they will feel bigger forces, and be more strongly driven to comply with changing resonance conditions. The piece(s) of cane constituting single and double reeds have small mass and relatively large areas in contact with the fluctuations, and they will respond promptly. The flute's air reed is even less massive. The sum of these characteristics produces a reed strongly influenced by pressure changes next to it. The player, except during the attack of a tone, and by his or her ability to change the air supply, has a less dominant role. On the other hand, the brass' massive and muscle-linked lip reeds are very much in the control of the musician.

The question remains, "How is the tone started? Everything you've seen so far started with the resonance already established in the cavity, with the highs and lows already traveling back and forth. What initiated these highs and lows, and at just the right frequency to be resonant? You know that the lows are created at the instrument's open end, and so the player doesn't have to make them directly. All he or she must do is produce the first high moving down the cavity and the cavity takes over from there. When the high, or low as you will see for air reeds, arrives again at the reed it inserts another puff into the cavity.

Free reeds are metal strips that vibrate according to their shape. They are not connected to resonant cavities, but are themselves springs that open whenever the force from the air source is greater than the spring's restoring force. The puff enters and passes directly into the outside air. The frequency of the metal strip's vibration is the frequency of the sound. The restoring force is zero when the reed is closed and increases as the spring opens, until it is enough to snap the reed closed again. Then the restoring force is zero again and the cycle repeats. The tone is started and ended by connecting and disconnecting the air supply. Chapter 10 describes free reed instruments.

Air reeds operate like railroad switches. A puff of air from the air supply is diverted into the instrument when a low arrives at the reed, changing that low into a high. Otherwise the air supply exhausts itself into the open atmosphere. This is probably why flute players complain that they have to keep blowing all the time.

Not so for the brass and their lip reeds. These reeds are massive and attached to muscles that the player commands. Except for the large mouthpieces of the bass brasses, the cross sectional area of the reed in contact with the cavity is rather small, and the pressure fluctuations in the cavity do not produce the big forces on the lips that would be needed to control them. The player is much more in command of what is happening in the cavity to the extent of being able to "lip in" tones not exactly the same pitch as the cavity's modes. As you will see in Chapter 8, "Brass Instruments," this ability to influence the pitch is important.

It is useless to study the action of a reed when it is not attached to the rest of its instrument. Without the resonant cavity's pressure fluctuations the reed will not operate the same way. Most cane reeds played by themselves sound like duck calls, not musical instruments, and air reeds disconnected from the instrument's cavity and free reeds disconnected from the air source don't make any sound at all. This is less true for the player controllable lip reeds. It is possible to play something of a tune on a trumpet mouthpiece alone.

7.5.1 Real Puffs

The previous graphs of pressure vs. time at the reed end of the instrument showed the puffs being added instantaneously, like this:

Chapter 7

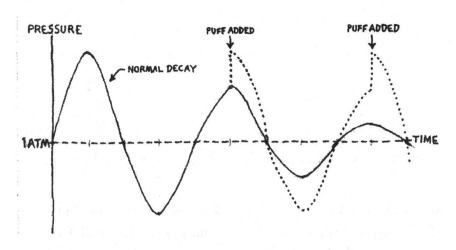

In reality the puff enters during a longer time interval.

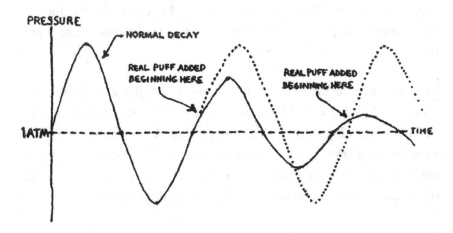

Just how long the reed opens the instrument to the air supply depends on the reed's inertia and the forces on it; and the finite duration of the puff can change the timbre of the tone. Here is an example.

Let's assume a conical cavity is trying to play its 2nd and 3rd modes together. The ratios of the frequencies of a cone are: 1 : 2 : 3 : 4 : ... etc., and we examine the 2 : 3 ratio pitches. The 3rd has less amplitude. Here is a pressure vs. time graph of the highs at the location of the reed.

Part 2: Musical Instruments

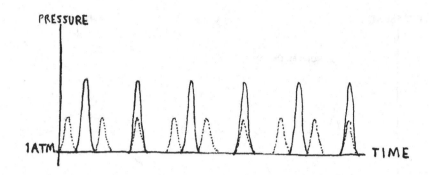

Suppose that only the higher amplitude pressure puffs of the 2nd mode will be large enough to open the reed and let in another puff. Even so, every so often the highs of the 2nd and 3rd modes happen at the same time, and part of the incoming puff will also support the 3rd mode. The louder 2nd mode gets a reinforcing puff every cycle, but the 3rd mode receives its puff only every third cycle. The 3rd mode here is less loud.

Note the three occurrences of the 3rd mode's highs straddling a 2nd mode's high. What would happen if the reed stays open longer causing the 2nd mode's high to become wider (taking more time)? What had been three separate highs now would tend to become a single, wide one, and the 3rd mode would lose its presence. In practical terms the 3rd mode would become less loud. This is a timbre change.

CHAPTER 8.
BRASS INSTRUMENTS

8.1 Ideal Conical Brass Instruments

The brass instruments play and sound the way they do because of two unique characteristics: their reeds are the players' lips and their shapes are combinations of cylinders and cones. Let's lay the groundwork for the study of these compromised devices by first studying an ideal and fictional completely conical brass instrument. Although at one time some brass instruments were wholly conical and had side holes like the woodwinds, that family has been abandoned. Brass now use slides or valves to change their lengths. Our idealized conical brass cannot have slides, but could have valves.

Begin by recalling that the frequencies of a cone's modes have ratios 1 : 2 : 3 : 4 : 5 : ... The musical intervals between these modes are OCTAVE, PERFECT FIFTH, PERFECT FOURTH, MAJOR THIRD, ... respectively. This is shown below with each line spanning an octave and the names of the necessary twelve semitones included.

do	di	re	ri	mi	fa	fi	sol	si	la	li	ti	do'

f | 11 semitones needed to play all the notes in this interval | $2f$

$2f$ | 6 needed for this P5 | $3f$ | 4 needed for this P4 | $4f$

$4f$ | 3 for this M3 | $5f$

Part 2: Musical Instruments

The frequency of the first mode is given the symbol, f. The number of semitone intervals between the first and the next to the last tones in the interval between the modes are also shown. These are the number of tubing length changes that must be made in order to sound all the semitones. For example, it takes eleven successive shortenings of the cone to play an ascending chromatic scale based on the 1st mode. While doing this, you would be playing the 1st mode of eleven shorter and shorter lengths of cone. It takes only six shortenings to span the 2nd and 3rd modes, and fewer and fewer to play between the higher modes. You could also begin at the higher mode and add length, lowering the pitch to span an interval.

Our "idealized conical brass" will have the same system for changing length as real brass: three keys that when pressed add length. Many real horns have three valves, which add length as follows:

- -- pressing the first valve adds enough tubing to lower the pitch two semitones.

- -- pressing the second valve adds enough tubing to lower the pitch one semitone.

- -- pressing the third valve adds enough tubing to lower the pitch three semitones.

All valves down produces a six semitone change. The trombone's slide has seven positions specified, but one is all the way in and corresponds to no valves pressed on the trumpet. Thus, both of these instruments can make six consecutive semitone changes. The conventional way to play these instruments is to be always adding tubing to the mode played. This causes a gap of five semitones between the 1st and 2nd modes, which cannot be played: di, re, ri, mi and fa. There are only six semitones between the 2nd and 3rd modes and from the 2nd mode on up all the semitones can be played using combinations of valves or the slide positions, some in more than one way.

Chapter 8

The problem of the gap of unplayable notes is solved in a most inelegant way. The octave between the 1st and 2nd modes is simply not used, and the lowest usable mode is the 2nd. The 1st mode or **pedal note** is discarded even though you can get down part way to it from the 2nd mode. The modes and semitones between them can be shown on a staff. Assume that the 2nd mode's pitch is middle C: C_4.

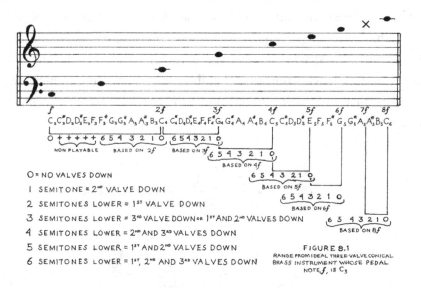

FIGURE 8.1
RANGE FROM IDEAL THREE-VALVE CONICAL BRASS INSTRUMENT WHOSE PEDAL NOTE f, IS C_3

Figure 8.1 shows that there are new possibilities for constructing tones. Any two numbers in a column, one below the other, indicate the two ways a note can be achieved. For example, the first semitone below the 3rd mode can be played by sounding the 3rd mode with the trumpet's second valve down, or by sounding the fourth mode with all valves down. Other tones that can be formed in more than one way are evident.

The pitch for the 7th mode is not a note on the staff. It lies between A_5 and $A_5\sharp$ and is drawn as a cross.

• QUESTION 8.1.   Calculate the frequency of $7f$ and then use Figure 2.2 to show that this is between A_5 and $A_5\sharp$.

Part 2: Musical Instruments

Calculate the frequencies of the six lower pitches you could get from $7f$ by using the valves. What are the nearest piano notes to these frequencies? •

8.2 Real Brass Instruments

Real brass are combinations of conical and cylindrical tubing. And, they have that flared end called the bell. The ideal conical brass explained above have the integer ratioed modes and some interesting possibilities for forming tones; also, they can play chromatic scales. Will combining cones and cylinders ruin this? Possibly yes. Then why are there real brass musical instruments? You know there are; let's examine them and find out why.

A typical trumpet is about half cylinder and half cone. The mouthpiece closes one end of the cylinder and the bell opens one end of the cone. Trombones are about one-third conical and horns are about one-third cylindrical. The half-and-half trumpet will be satisfactory for analysis. Let's look at the bad news about the loss of even tempering first.

Chapter 8

What are the frequencies of the modes of our cylinder-cone? As you expect they will be somewhere between those for a pure cone and for a pure closed-end cylinder. And yes, this means that the modes' frequencies are not integers times the 1st mode's frequency. You've found that the modal frequencies of a closed-end cylinder are

$$f_{cyl} = ((345 \text{ m/s}) / 4L) \, n, \qquad n = 1, 3, 5, \ldots \qquad (7.1)$$

and those for a cone are

$$f_{cone} = ((345 \text{ m/s}) / 2L) \, n, \qquad n = 1, 2, 3, \ldots \qquad (7.2)$$

For our trumpet, both the cylinder's length and the cone's length are the same, L. Therefore for the 1st mode, where $n = 1$,

$$2(f_{cyl}) = f_{cone}.$$

For the 2nd mode, $n = 3$ for the cylinder and $n = 2$ for the cone,

$$(4/3) f_{cyl} = f_{cone}.$$

• QUESTION 8.2. Start with the formulas for the modes' frequencies of cylinders and cones,

$$f_{cyl} = ((345 \text{ m/s}) / 4L) \, n, \qquad n = 1, 3, 5, \ldots$$

and,

$$f_{cone} = ((345 \text{ m/s}) / 2L) \, n, \qquad n = 1, 2, 3, \ldots ;$$

and show that for equal lengths of cone and cylinder the frequencies of their 1st modes have the relation,

$$2(f_{cyl}) = f_{cone}. \bullet$$

Part 2: Musical Instruments

- QUESTION 8.3.   Start with the formulas for the modal frequencies of cylinders and cones, and show that equal lengths playing their 3rd modes will have frequencies with the relation,

$$(12/10) f_{cyl} = f_{cone} \cdot \bullet$$

- QUESTION 8.4.   Examine the formulas for the relations between the frequencies of the same modes for equal lengths of cone and cylinder; i.e., look at

$$2(f_{cyl}) = f_{cone}, \text{ for the 1st mode}$$

$$(4/3) f_{cyl} = f_{cone}, \text{ for the 2nd mode}$$

$$(12/10) f_{cyl} = f_{cone}, \text{ for the 3rd mode.}$$

What is happening to the differences between the cone's and the cylinder's frequencies for the same mode as the mode increases? •

With our usual simplicity pick the cylinder's 1st mode frequency to be 100 Hz, and use (7.1) and (7.2) to calculate the modes' frequencies for equal length closed-end cylinders and cones. The results, along with a suggested real frequency for our composite trumpet are shown below.

MODE	1st	2nd	3rd	4th	5th	6th	7th	8th	9th
CONE, Hz	200	400	600	800	1000	1200	1400	1600	1800
CYL, Hz	100	300	500	700	900	1100	1300	1500	1700
REAL, Hz	150	350	550	750	950	1150	1350	1550	1750

not used   OCTAVE +50Hz   OCTAVE+50Hz

750/350=2.14   1550/750=2.07

Chapter 8

The real frequencies were calculated using the assumption that they are half way between those of the cone and cylinder. The 1st mode is not used for the same reason as before. Each real "octave" is seen to be 50 Hz too sharp. Their musical octave interval frequency ratios are 2.14 and 2.07 instead of the wanted 2.00. This is not a good situation, but also not very bad. And, it appears that it is improving for the higher pitches.

• QUESTION 8.5.   Calculate the musical octave interval frequency ratio for the next octave (not shown) in the table above. Is it true that the higher and higher octave intervals will be less and less sharp? Why?
•

Still, is there any way this sharpness of the octave intervals could be corrected? Of course, trombone players could do this by moving the slide a bit, but horn and trumpet players are stuck with the fixed action of the valves. Could there be some way to lower the frequency of the higher notes in the interval or possibly raise the frequencies of the lower ones? Making the cavity longer for the higher notes or shorter for the lower ones would do this. The bell, as you will see, does the former. It also produces less and less increase in the cavity's length as the frequencies increase. This means that the alternate valve combinations shown in Figure 8.1 would still be available for our half-cone and half-cylinder instrument; and this is almost too good to be true.

8.2.1 The Bell

In addition to providing the more gradual transition to the outside air mentioned in Section 7.4.3, the bell changes the effective length of the resonant cavity. I stated that the cavity should be lengthened for higher pitches, and this is just what the bell does. In order to understand the bell's contribution to the length of the cavity we must examine again the reflection that takes place at the open end of a cavity. In particular we must return to the question of the difference between a cone and a bell. There is a rule of reflection that is applicable at both the open and closed end of a cavity: the sound wave will be at least partially reflected at places along the cavity where the pressure abruptly changes. In practice this happens wherever the cross sectional area of the cavity suddenly changes, forcing the traveling highs and lows to quickly change volume. This rule is easy to apply to a closed end; the cross sectional area suddenly becomes zero and the whole wave has to compress there and bounce back: total reflection. At a sudden open end the cross sectional area becomes infinite; this is certainly a sudden change. The highs and lows meet atmospheric pressure there, and part of the wave leaves the cavity and part changes from high to low, or vice versa, and this reflects back into the cavity to establish the standing wave and to control the reed. So, cones must change shape slowly enough to prevent reflections except where needed at their open ends.

What is an abrupt change of cross sectional area? Theory says that it happens wherever the cavity's cross sectional area changes a lot within the distance of a wavelength. This is not an exact definition, but it will do here.

Now let's see how this theory is applied to a bell.

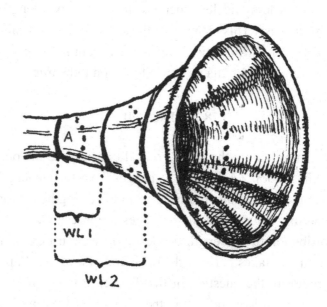

Chapter 8

First note that the bell has been flared out. Its cross sectional area increases much faster toward its open end. Study the sketch and note how the bell's cross sectional area increases, starting from area A, for two wavelengths $WL1$ and $WL2$, about an octave apart. Shorter wavelengths indicate higher frequencies and vice versa. So, we want the action of the bell to lower the higher frequency (thus lengthen the cavity for $WL1$).

Wave 1 enters the bell through area A; and after traveling a distance equal to its wavelength ($WL1$) the bell's cross sectional area has not changed much. So, it keeps traveling until it reaches a place where the bell is more flared. Wave 2 also enters the bell through area A, but at a farther distance ($WL2$) down the bell the cross sectional area has changed a lot. This, according to our theory, causes wave 2 to reflect at about area A, while wave 1 continues on before it reflects. The bell is longer for wave 1 than for wave 2. Just what we need. The bell is not only an efficient radiator of the sound, but creates a longer cavity for higher frequency waves. It is truly an amazing device, and much older than the scientific explanation of its operation. A genuinely quantitative description and explanation of the operation of the bell is not suitable for this book[8].

• QUESTION 8.6.   Measure the radii of the cross sectional areas at A and at distances $WL1$ and $WL2$ down stream from A. Then calculate the values of these cross sectional areas (area = πr^2) at these three places. Finally calculate the percentage increase of each of the two cross sectional areas compared to the area at A. From this information argue that wave 1 will reflect farther down stream in the bell than wave 2. •

The loudness of brass instruments indicates that not much of the sound wave is reflected back down the cavity. This causes a weaker standing wave in the cavity, which in conjunction with the direct muscular control between the player and the lip-reed, allows the musician an additional control of the pitch.

There have been wild designs made to overcome the need for "lipping in" the pitch. Figure 8.2 shows one by the otherwise praised inventor of the saxophone: Adolphe Sax. It is a six valved instrument which is actually seven instruments with modes a semitone apart played one at a time by pushing, or not, one valve. It weighed too much.

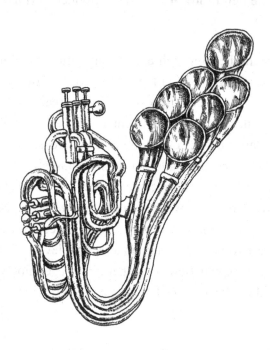

Figure 8.2 A Type of Tenor Trombone

Sax also tried this configuration for a cornet.

CHAPTER 9.
WOODWINDS

9.1 Ideal and Real Woodwinds

SOME WOODWINDS

SHOWN TO SCALE FROM LEFT TO RIGHT:
BASSOON, FLUTE, OBOE, SAXOPHONE
CLARINET, RECORDER

No separate discussions of "ideal" and "real" woodwinds are needed. Real woodwinds are almost perfect. They are conical, or open at both ends, which is acoustically the same thing (oboe, bassoon, saxophone, recorder, flute) or cylindrical and closed at one end (clarinet). These characteristics set the ratios of the modes' frequencies as ratios of integers. This does leave a big frequency gap between the clarinet's first two modes (1 : 3), whereas the conical instruments have an octave (1 :2) here. Nevertheless the first mode is used, and many side holes are present along the bore to shorten its length as they are opened from the bottom up to connect the resonant cavity to the outside air. The reed is very much under the control of the cavity's pressure oscillations, and mechanical devices are used to change modes.

With its keys removed, the conical shape of a saxophone is obvious. The side holes, also called tone holes, get bigger as the cavity's bore increases. This is necessary if the sound wave is to think that the cavity ends at the open tone hole. Not all the sound leaves the instrument through the end of the cone; some exits through the open downstream tone holes. They must be big enough to create a big change in the cavity's cross sectional area when they are uncovered. This fools the wave into thinking that the location of the open tone hole is the end of the cavity; therefore, an open-end reflection happens there. As far as the wave is concerned, the cavity has a new length and the frequency changes to be resonant for it, and the pitch changes. A bell is neither needed nor present.

Chapter 9

DEMONSTRATION XXI: Open Side Holes Change Length of Cavity

Apparatus: Bass Recorder

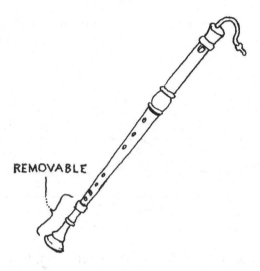

When playing notes that leave most of the bottom holes uncovered, the whole end can be removed and the pitch barely changes. The resonance in the bore believes that the instrument ends near the lowest open hole. Listen to the recording and note the slight change in pitch when a sizable part of the cavity is removed (Band 22).

The problem of changing from mode to mode while playing is solved by adding a key (or keys) which, when open, tends to clamp its location along the bore at atmospheric pressure. This suggests to the resonant standing wave that it should have a node there. And the key is placed so that this node is one for the desired mode. Here are pressure vs. distance graphs for the first two modes in a cylinder of length L and open at both ends. Recall that the boundary conditions for this cavity specify atmospheric pressure at both ends.

Part 2: Musical Instruments

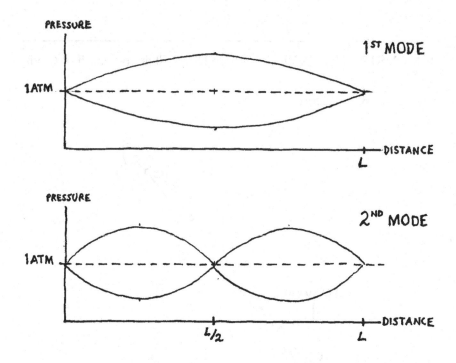

A small, keyed opening at $L/2$ would, when open, tend to clamp this location at the atmospheric pressure needed by the 2nd mode. This key can't be too big, or the cavity will think it ends there. This will cause a shorter resonant length cavity for the 2nd mode and the side holes downstream of this key will be useless. These small holes are called **speaker holes** or **register keys**.

• QUESTION 9.1.   Sketch graphs such as shown above for the first two modes of a clarinet (cylinder closed at one end). Then show that its register key is not positioned halfway down the clarinet's acoustical length, L. Where is it?

Look at the clarinet's range of pitches in Figure 2.2. How could one register key be enough? Clarinet players will recognize that this is a trick question. Read the answer proposed in the Answers to the QUESTIONS section •

CHAPTER 10.
TWO OTHER KINDS OF WIND INSTRUMENTS

Harmonicas, accordions, and concertinas; ocarinas and sweet potatoes; it would be easy to skip them. But they are somewhat popular, and probably more importantly, they don't function like the instruments you've read about so far. Of course, they produce musical sound waves, but each of these two groups has a unique way of doing so.

10.1 Harmonicas, Accordions, Concertinas

HARMONICA, CONCERTINA, ACCORDIAN

The harmonica, accordion, and concertina do not use thin strings or resonant cavities to specify the pitch. Indeed, they are too small to contain the necessary lengths. They do contain spring-steel reeds, one for each pitch, which are filed to make them more flexible or weighted on their flapping ends to make them vibrate more slowly. In either case each vibrates with the frequency of its pitch. The vibrations open and shut little doors through which the air puffs exit into the outside air, and these puffs are the sound waves you hear. A bellows (accordion or concertina) or the player's lungs (harmonica) pump or draw the air through the reeds. Here is a schematic diagram of this arrangement for the accordion or concertina.

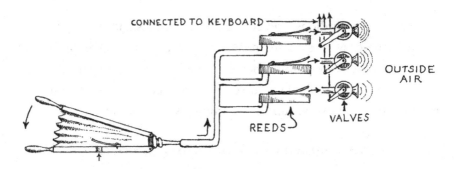

Pressing a key opens a valve, and lets air flow through the particular reed. When the bellows is compressed, its volume decreases and the pressure increases. When it expands, it creates lower pressure. So, air can flow in either direction through the reed. Pairs of reeds, each having the same frequency, or not, are screwed on each side of metal blocks against rectangular holes as shown below.

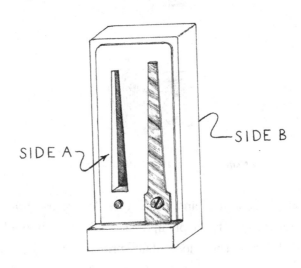

Chapter 10

One side of the block is connected to the bellows, the other to a valve and then to the outside air. When a key is pressed, its valve opens, and higher pressure on side A closes the reed on side A and opens the reed on side B. If the pressure is not so great that it defeats the restoring action of the reed on side B, this reed will vibrate. If the pressure is lower on side A, the normal air pressure on side B will open the reed on side A and start it vibrating. Compressing and expanding the bellows cause these higher and lower pressures.

The harmonica player blows or sucks, and opens and closes the air supply to the reeds with his or her mouth or tongue. There are no keys and thus, the harmonica's "valves" are upstream of the reeds.

10.2 Ocarina

The ocarina neither looks like a "little goose" nor like any instrument we've seen so far.

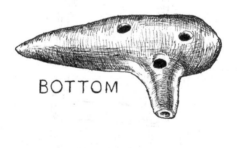

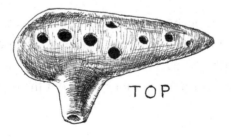

To play it you blow softly and uncover holes to raise the pitch. Nothing is hidden inside; its fipple makes it an air reed instrument, but there is no obvious open end anywhere else and the finger holes don't make the interior cavity longer or shorter. Opening the holes in any order will raise the pitch. How does it play pitches?

It is an example of a **Helmholtz resonator**[9], and like many musical instruments is older than the explanation of how it works. Let's begin our explanation with two demonstrations of Helmholtz resonators.

DEMONSTRATION XXII: Resonance in a Bottle

Apparatus: Bottle with Finger-sized Neck Opening

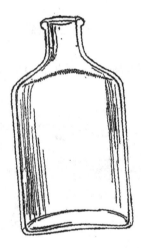

Put your finger in the neck of the bottle and snap it out. A tone is heard. It lasts slightly longer then the time it takes to snap out your finger, which shows that high and low pressure vibrations are happening in the bottle. A portion of the vibrations leaks out the neck and you hear this sound.

Substituting a bigger bottle with the same size neck produces a lower pitch tone.

Chapter 10

These are not sustained sounds; there is no reed to replenish the puffs. But, the pitch does indicate that there was a resonance in the bottle. Don't be in too much of a hurry to conclude that the bottle is just a strangely shaped open-ended cylinder. The next demonstration shows that it is not.

DEMONSTRATION XXIII: A Blown Helmholtz Resonator

Apparatus: Used Cartridge Casing

The casing is cut open at both ends.

Put your thumb over one end and blow across the other to make a whistling sound. Turn the casing over and repeat. You get a different pitch now. Same length casing, but different pitches.

Here is a model that explains what was happening during these two demonstrations. The compressible air in the body of the bottle and casing acts like a spring. The air in the neck acts like a bob. The bottle and casing are pneumatic equivalents of the spring-bob vibrator you saw in DEMONSTRATION XIX in Section 7.3. The one below has no "rubber band chain" to sustain the oscillations.

Part 2: Musical Instruments

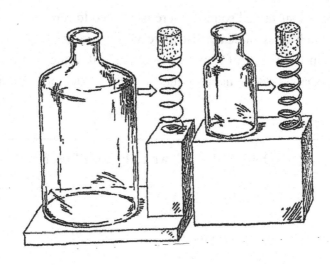

If you used a bigger bottle there would be more air in the body to compress, and it would take a bigger compression to make the same pressure change. This corresponds to a less stiff spring and causes a lower natural frequency of vibration. This explains the lower pitch from the bigger bodied bottle, and indeed, bigger ocarinas have lower pitches. Turning the cartridge casing over exchanges the volumes acting as bob and spring and the pitch changes too. This is evidence that the spring-bob model works for a Helmholtz resonator.

The ocarina is made of inflexible clay, which fixes its volume. According to our spring-bob model, this also fixes the stiffness of the spring. Opening the holes raises the pitch; this must be equivalent to decreasing the mass of the bob. How else can you increase the frequency of a spring-bob system if the spring is kept the same?

But, here, our spring-bob model fails a bit, also. Opening more holes would seem to be adding more bob. This would decrease the frequency of a spring-bob's oscillations, but instead it raises the ocarina's pitch. What opening the holes does do is increase the total area connecting the interior volume to the outside air. This lets the high pressure inside the ocarina relieve itself quicker, and the next puff enters sooner through the fipple. This is a frequency increase.

10.2.1 Helmholtz Resonators in Other Woodwinds

Look again at the stripped saxophone in Ch. 9, and note that although the wall of the sax is thin sheet metal, its side holes have short lengths of tubing connecting them to the outside air. These small cylindrical volumes and also those in the thick-walled wooden woodwinds cause small modifications to the pitch, which is still mainly determined by the cavity's length. The air in these small cylindrical volumes becomes miniature bobs of a spring-bob resonating system when the side holes are open, and add volume to the main cavity when they are closed.

A thorough explanation of how these short pieces of tubing change the cavity's resonant frequencies is beyond the objectives of this book. See Chapters 21 and 22 in Arthur Benade's *Fundamentals of Musical Acoustics* (recommended in the Additional and Extended Reading section) if you want to know more.

It is sufficient here to note that these small pieces of tubing lower the resonant frequencies of a long cavity whether or not the side holes are open or closed. The amount of lowering is roughly proportional to the ratio of a tube's volume to the volume of the part of the main cavity near the tube's side hole.

As would be expected the holes are most effective this way when they are located at antinodes along the bore. At nodes there are no pressure fluctuations present to be affected. If an open side hole has a big enough diameter to convince the sound inside the resonant cavity that the cavity ends there (as it should), the sound will exit there and the remaining downstream side holes will have little effect.

There is a tale about recorder maker Arnold Dolmetsch (1858-1940) tuning his customers' instruments with a pocket knife while they nervously watched. Even if Dolmetsch did not understand the mathematical descriptions of sound such as Helmholtz', he got the recorders tuned.

The idea of artisans perfecting techniques or devices prior to there being scientific descriptions of them has been noted by Cyril Stanley Smith (1903-1992), metallurgist and historian of science and technology. Smith argued that much of our science and technology has artistic roots. Musical instruments are an example; certainly many were in good use before anyone really understood how they worked. In fact, other than explanations of broad fundamentals, such as presented in this book, the reasons are often still not clear.

• QUESTION 10.1.   Locate a biography of Cyril Stanley Smith and discuss what, in your opinion, was the most important metal he worked with. •

• QUESTION 10.2.   Locate a biography of Arnold Dolmetsch and discuss how such a comparatively recent person acquired an interest in baroque and renaissance instruments such as the recorder, which had generally been replaced by the transverse flute by about 1750. •

PART 3: ACOUSTICS

CHAPTER 11.
ROOM ACOUSTICS

Building an auditorium or hall seems to be a two-stage work: first, the room is designed and built, and then several years later it is modified to try to eliminate its acoustic problems[10]. Why are these two steps necessary; why can't the room be right at first? Right for what, though?

Up to now this book has almost exclusively been describing sound in or very near its source. Room acoustics describes sound at the location of the listener.

Once it was enough that the audience just heard the music. Early opera halls used every bit of space to pack the people in close to the source. Even the walls disappeared in favor of boxes. Figure 11.1, the original Teatro alla Scala in Milan looking out into the audience from the back of the stage shows this.

Part 3: Acoustics

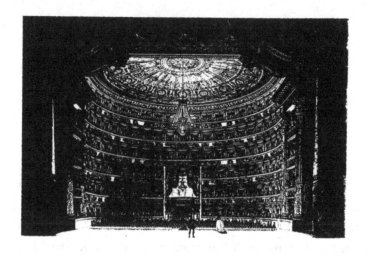

Figure 11.1 The Original Teatro alla Scala

A quite different approach used the walls as hard surfaces to reflect the sound throughout the room. Figure 11.2, St. Martin's Hall in London shows an example. It also indicates that not all the audience was intent on the music. No wonder, the acoustics of such a shoebox must have been awful.

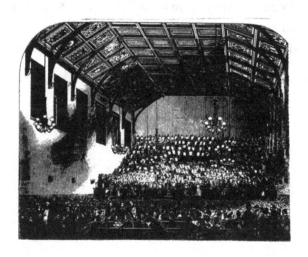

Figure 11.2 St. Martin's Hall

Chapter 11

The general rule of thumb seemed to be that if everybody could see the musicians at least some of the sound would come directly to them. This was, and still is, not a bad idea.

11.1 The Language of Musical Acoustics

11.1.1 Musical Descriptions

It's no longer enough that the listeners just be able to hear the music. Now, good sound is expected, exhibiting the proper characteristics of the music and its performance. These will be different for music by large orchestras, smaller groups, and soloists, and this is why it is a difficult problem to achieve good acoustics in a modern multipurpose room. Well then, what is good sound?

What do the terms "clarity, warmth, dryness, top, bottom, liveliness, intimacy, dynamic range" and others used to describe how the music sounds mean? These words have common use but not common meaning. Attempts by scientists to define them produce technical language, which, for most listeners, is neither understood nor appreciated. A problem with these terms is that they are listeners' perceptions, and it is possible that listener A will say a sound lacks warmth while B claims that sound is too warm. Sometimes it is not even clear whether the sound source or the room is making these defects. Until listeners agree on the nature of the problem, and perhaps they won't, there is no universal acoustic solution.

You already know that differences in each listener's ear-brain system will cause him or her to hear a sound differently, and so even professional listeners may not agree that the sound has any particular characteristic. Also, a room that lacks warmth for symphonic music may sound just right for small groups or individual recitals. The room isn't changing the sound; it's just that most listeners want their symphonies with a warmer sound.

Part 3: Acoustics

Good acoustics are what make the music sound good to the listeners, and consist of whatever it takes to do this. Scientifically perfect acoustics are not the answer. Some of the terms, such as "dry, lacks warmth, missing top or missing bottom, lack of liveliness" used to describe perceived defects may indeed be describing the need for less perfect physical acoustics, while others such as the room having "dead spots" or echoes indicate just the opposite, that the physical acoustics need to be improved.

Manfred Schroeder in the Anechoic Chamber at Bell Laboratories

Manfred Schroeder[11] sidestepped the problem of trying to assign quantitative values to these properties, and simply played the same music in different rooms and asked people in which room it sounded best. The rooms had various values of measurable characteristics, and he was able to find a consensus about what constituted good acoustics as far as these characteristics were concerned. He did not find universal agreement, only a consensus. Trying to improve acoustics often becomes trying to impose a scientific solution on an ill-defined problem, an uncertain process.

And don't forget that all the listeners are not in the audience. The musicians must hear each other, too, if they are to play together. They must hear their comrades directly, but also what the audience is hearing: sound in two very different places. The word **ensemble** describes how well the musicians hear each other; how well they know what the audience is hearing is a more complicated question, especially if the sound is different throughout the room.

11.1.2 Scientific Descriptions

Whether or not agreement about what is needed is reached, there are measurable properties of a room that can be changed to modify the subjective acoustics. These are: **resonances** in the room, **reverberation time,** and **delay time.** These three all depend on how the sound reflects from the holes, obstacles, and the boundaries of the room. Sections 7.1, 7.2, and 7.4 introduced you to sound **reflections** from open and closed ends of cavities. These results apply to the open and closed ends of rooms, too. What you have not yet discovered so far is what happens to the sound wave if it encounters a normal size object in the room. If this object, a hole or an obstacle, is about the same size as the sound's wavelength, a spreading and redirection of the wave occurs called **diffraction.** And, in addition to reflection and diffraction, any time a wave hits something the amplitude of its pressure fluctuations may diminish. This is called **absorption.** Appendix A. "Sound Becomes Less Loud" describes reflection, diffraction, and absorption as well as the natural decay that occurs when the sound simply moves away from its source. If you can control reflection, diffraction, and absorption you will have powerful tools to control resonances, reverberation time, and delay time, too. Doing this, and making sure that the loudness is satisfactory throughout the audience, you can give a room almost any acoustics you want. So, let's take a careful look at each.

Part 3: Acoustics

11.1.2.1 Reflection and Resonance

Begin with a review of standing waves, nodes, and antinodes, especially in those cavities we call rooms.

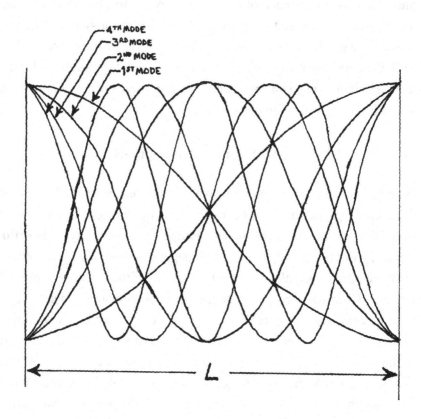

Figure 11.3 Standing Waves Between Walls a Distance L Apart

Figure 11.3 shows the envelopes and the locations of the pressure nodes and antinodes for the first four modes of standing waves between walls a distance L apart. The high and low pressure puffs of the compressible air hit the walls and bounce back just like an elastic rubber ball. Thus, the walls are the locations of pressure antinodes and motion nodes. These are the kinds of resonances you would get between solid walls. If one of the walls were replaced by an opening (openings, whether in cylinders, cones, or rooms cause some reflection), it would then be the location of a pressure node, and the type of resonances and standing waves you found for open-ended cylinders would occur. Pairs of adjacent nodes, or antinodes, will be 1/2 wavelength apart; and a mode's adjacent nodes and antinodes will be 1/4 wavelength apart. These spacings also are shown in Figure 11.3. Note that a mode's nodes and antinodes become closer together as its frequency increases.

Recall that although strings or cavities have resonant frequencies determined by their length, it is the reflections at the boundaries that supply the second wave that superpositions with the original one to form the standing wave. If a room has the right length, and also is not too big, the reflections from opposite walls can combine to make resonant standing waves across the room. If, on the other hand, the room is big, or if its walls are poor reflectors, the reflected wave from one wall will lose most of its amplitude before it returns very far into the room, and the room will act like it has only one wall. Near that wall the reflected wave will interfere with the original wave to produce nodes and antinodes, but they are not really standing waves or resonances because there is only one reflecting boundary. We could call these "nodes and antinodes near a wall" or something else. They have no real names, so let's give them a shorter one: **near wall effects**.

Rooms by themselves are not musical instruments and cannot create sound. They can modify it, however. A room, just like a wind instrument, has its resonant modes determined by its length. Unlike the instrument's length, the room's length doesn't change; however, there are three lengths, each with its own set of resonant frequencies, and six walls (including ceiling and floor) for creating near wall effects. If one of the lengths happens to be correct for a resonance with one or more of the music's tones, and if the boundaries of the room are good reflectors, the room will "play" this resonance. Even in a large room having reflecting boundaries far apart, near wall effects could cause nodes and antinodes of loudness near the reflecting walls, and listeners at these nodes and antinodes probably will encounter acoustic problems. So, these possible acoustical defects will depend on both the room and the music played in it. Any fixed room design can have problems.

Will there be a problem in rooms because some notes are resonant and others not? As usual, in this world, it depends. The dimensions of the room will determine the extent of the nodes and antinodes, and thus, whether it is possible for a listener to be at one. If so, he or she will hear that frequency too softly or too loudly, or as you will see, with some other distortion of the sound. Let's look at a rather long example of this.

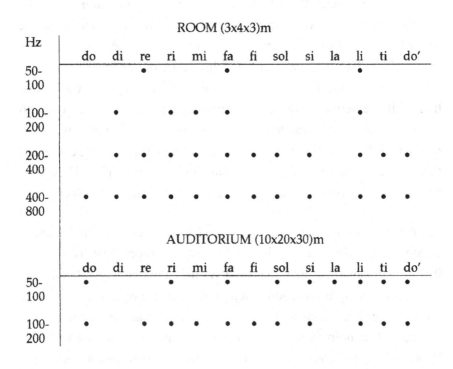

Table 11.1

Resonant Modes in a Room and in an Auditorium

Table 11.1 shows the resonant frequencies for a house-size room (3x4x5 m) and an auditorium (10x20x30 m). The actual frequencies, although

Chapter 11

not listed, were calculated exactly the same way as for cylinders closed at both ends using equation (7.2):

$$f_n = ((345 \text{m/s}) / 2L) \, n, \qquad n=1, 2, 3, \ldots \qquad (7.2)$$

Now, L is the length, width, or height of the room or auditorium. The dots indicate the resonant frequencies and are placed below the notes that they represent or are near on a musical scale. The scales are divided into octaves starting at the just audible 50 Hz.

Notice how many resonant tones there are and how they are spaced throughout the scales. The smaller room will definitely have both resonant (loud antinodes and dead nodes) and non-resonant notes in the first two octaves of audible tones. The room will not treat equally tones whose frequencies are below about 200 Hz. Beginning in the 3rd octave (200-400 Hz) the notes will be, more or less, all resonant.

In the auditorium, with its longer dimensions, the resonant frequencies of the lower modes' tones are too small to be audible. For example, I calculated that its 30 m dimension has a first resonant mode of 5.8 Hz. Its 9th mode, 52 Hz, will be the first one that you can hear, and then the resonances will be spaced 5.8 Hz apart. Except for the lowest notes in the 50-100 Hz octave, all the audible notes will be resonances.

Resonant notes will have nodes and antinodes spaced throughout the room, but the non-resonant notes will not, and will sound about the same everywhere. So will the resonant ones, provided that the listener is not sitting at the location of a node or an antinode. Figure 11.4 shows a perplexed listener with the nodes and antinodes for three unfortunate situations.

Part 3: Acoustics

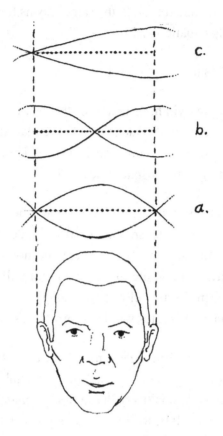

Figure 11.4 Listener and Standing Waves

In case a.) the listener has both ears at nodes and will not hear that frequency. In case b.) he or she will hear it too loudly. In case c.) the person's right ear, at a node, will hear nothing while the left, at an antinode, hears the tone too loudly. The listener in this case might think that the tone is coming only from the left. These misfortunes can also happen to listeners fairly close to walls where all frequencies produce near wall effects. Let's calculate the range of wavelengths that by being resonant or causing near wall effects could produce these aural distortions.

Chapter 11

In the equations that follow it will be awkward to continue writing WL for wavelength. There is a single symbol for this: it's the Greek letter lambda, λ. For example, equation (2.3) was written as,

$$v = (WL)f.$$

With the symbol λ it becomes,

$$v = \lambda f. \tag{11.1}$$

Subscripts will continue to be used for the mode number where needed: WL_1 becomes λ_1, and so forth.

A typical listener has two ears separated by about 0.18m. Will problems occur if the distance between a resonant frequency's nodes and antinodes, or between nodes or antinodes, gets to be this 0.18m? First, let's find out which wavelengths and frequencies could cause these problems.

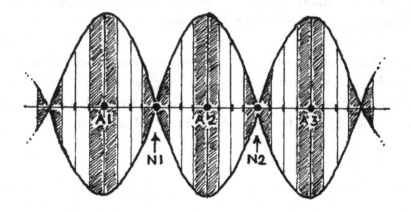

Figure 11.5 Standing Waves Divided into $\lambda/16$ Wide Regions

Figure 11.5 shows part of a standing wave divided into shaded and unshaded regions. I have divided each half wavelength, between nodes, into eight equal lengths, each of these $\lambda/16$ long, and have shaded the first, fourth, fifth, and eighth ones. An ear located at a shaded region will hear the tone too softly (in the 1st, and 8th regions where the pressure fluctuations are small), or too loudly (in the 4th and 5th regions where they are big). Loudness will be about normal in the unshaded regions.

This division is rather arbitrary. It will make the following analysis easier, and will cause the loudness of the sound within the white stripes along the distance axis to be about 15% and 85% of the maximum loudness at the nodes and antinodes at N1 and N2, and A1 and A2 and A3. This is my choice for sound that is neither too soft nor too loud.

• QUESTION 11.1.   Loudness of sound depends on the square of its amplitude. Nodes are locations of zero amplitude; there is no loudness there. Antinodes are at the locations of maximum amplitude, and loudness is maximum at their locations.

Show why the loudness at the locations of the white stripes will be about 15% and 85% of the maximum loudness. The solution to this QUESTION requires you to use the advanced trigonometry of sines and cosines. Therefore, this QUESTION is optional. •

So, if a listener has one ear in the shaded region around N1 and the other in the shaded region around N2, he or she will be a more-or-less satisfying case a.); and will hear that tone too softly or not at all. The *shortest* wavelength for this is shown in the below sketch.

Chapter 11

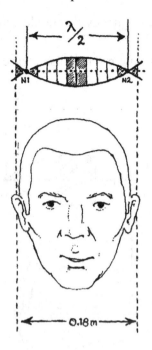

This sketch shows 10 of the stripes in the distance between the ears. Each stripe is $\lambda/16$ wide. Note that $\lambda/2 = (8/16)\lambda$. So,

$$10(\lambda/16) = 0.18 \text{ m, or that}$$

$$\lambda = 0.18 \text{ m} \, (16/10) = (8/5) \, 0.18 \text{ m} = 0.29 \text{ m}.$$

Note that the above equation can be written as,

$$\lambda/2 + \lambda/8 = 0.18 \text{ m}.$$

We will use this form of the equation below.

The *shortest* wavelength for case a.) has the frequency using equation (11.2)

$$f = v/\lambda = \frac{345 \text{ m/s}}{0.29 \text{ m}} = 1190/\text{s}.$$

Part 3: Acoustics

So 1190/s is the *highest* frequency that could cause a case a.) problem.

The *longest* wavelength that could cause a case a.) problem is shown in the sketch below

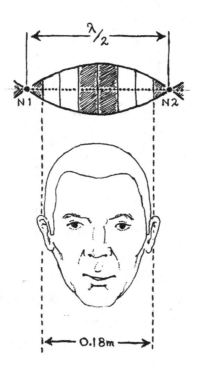

Now only 6 stripes equal 0.18 m. Or,

$$6(\lambda/16) = 0.18 \text{ m, and } \lambda = 0.48 \text{ m.}$$

This equation can be written as

$$\lambda/2 - \lambda/8 = 0.18 \text{ m,}$$

and both the case a.) wavelengths can be calculated from

$$\lambda/2 \pm \lambda/8 = 0.18 \text{ m.} \tag{11.2}$$

Chapter 11

The ± symbol is called "plus or minus". Equation (11.2) calculates two values of λ, one when the plus sign is used, and the other when the minus is used. The values of the wavelengths that can cause the listener to hear sounds too softly are these two wavelengths and all of those in between them.

The frequency of the *longest* wavelength is

$$f = \frac{345 \text{ m/s}}{0.48 \text{ m}} = 720/\text{s}$$

and the range of case a.) problem frequencies is

$$720/\text{s} \leq f \leq 1190/\text{s}.$$

The symbol ≤ reads "is less than or equal to". The symbol ≥ reads, "is greater than or equal to".

Case b.) can be analyzed using the same methods except that now $\lambda/2$ is the distance between A1 and A2 in Figure 11.5.

The results are the same as found in case a.), and the formula

$$\lambda/2 \pm \lambda/8 = 0.18 \text{ m}$$

will also calculate the longest and shortest wavelengths for case b.). The problem-causing frequencies are the same, too:

$$720/\text{s} \leq f \leq 1190/\text{s}.$$

Figure 2.2 shows that most instruments play in this range of frequencies, and thus, standing waves in an auditorium or hall are possible and could cause distorted listening.

Case c.) problem wavelengths and frequencies are calculated using the same methods, except the distance between the listener's ears is now from N1 to A2, plus or minus some stripes as in case a.). The troublesome wavelengths and frequencies are found from

$$\lambda/4 \pm \lambda/8 = 0.18 \text{ m} \tag{11.3}$$

and are 0.48 m and 1.44 m; and the frequency range is

$$240/\text{s} \le f \le 720/\text{s}..$$

It appears that loudness problems (too much or a lack of in one or both ears) can be caused by standing waves for frequencies between 240/s and 1190/s

• QUESTION 11.2. Calculate the troublesome wavelengths and frequencies for case b.). Or, if you prefer, explain in a paragraph why these are the same as for case a.) •

• QUESTION 11.3. Calculate the troublesome wavelengths and frequencies for case c.). •

Table 11.2 shows the range of wavelengths and frequencies which might cause bad seats in music halls, either due to near wall effects, or if the dimensions of the room make it resonant for any of these frequencies.

CASE	λ, m	f, Hz
a.), b.)	$0.29 \le \lambda \le 0.48$	$1190 \ge f \ge 720$
c.)	$0.48 \le \lambda \le 1.44$	$720 \ge f \ge 240$

Table 11.2 Wavelengths and Frequencies that Could Cause Bad Seats

In all cases, however, if the listener moves to the right or left enough so that the ears are no longer near nodes or antinodes the tone will sound fine. How far must he or she move? A listener can move out of a bad

Chapter 11

location by moving a distance less than $\lambda/8$, which will take the ears out of a shaded region and into a clear one where the sound is much less uneven. Even for the longest wavelength in Table 11.2 for cases a.) and b.), this is

$$0.48 \text{ m} / 8 = 0.06 \text{ m}.$$

A small motion of the head would do it, but it took a bit of analysis to find this out. Using the same method for case c.) shows it might take as much as a neck cramping 0.18 m to correct for sound apparently coming from the wrong direction.

In addition, for low frequencies it is possible to have both ears in the *same* shaded region and hear resonant tones too loud or too soft. Examine Figure 11.5. to see that this is possible if

$$0.18 \text{ m} \leq \lambda/8, \text{ or } \lambda \leq (0.18 \text{ m})8 = 1.44 \text{ m}.$$

A wavelength of 1.44 m corresponds to a frequency of 240 Hz. So, resonant frequencies or near wall effects due to frequencies below 240 Hz can cause this problem. These potentially troublesome frequencies are the important low tones. This presents a real problem because the listener might have to move as much as $\lambda/8$ to get both ears out of the bad region. This is at least 0.18 m, and for the lowest audible frequency, 50/s (wavelength of 6.9 m), would be 0.86 m. In order to miss the bad seats the audience would have to be rather widely spread out in the room. Here ends the example begun just before Table 11.1 in Chapter 11. Each size room must be analyzed accordingly. There is no single, universal fix.

However, it is almost certain that the room will not have a simple, hollow, rectangular shape; and thus, the analysis becomes even more complicated. Too much so to be done by a hand calculation; a high-speed computer running sophisticated software is necessary. Or, on the other hand, you could make a direct measurement of the sound in a scale model of the room. Of course, the sound's frequency must be scaled up, too.

• QUESTION 11.4.   Explain why a listener could have both ears in the same node or antinode, if one eighth of the resonant wavelength were at least 0.18m. •

• QUESTION 11.5.   A listener has both ears in the same node or antinode. Explain why he or she must move at least 0.18m in order to hear the tone with normal loudness. •

Since acoustical problems are caused by resonances, why not try to get rid of them? Because resonances and near wall effects are caused by reflections, let's see if we can reduce the walls' ability to reflect. Whether we want to or not is another question. As you will shortly see the reverberation time depends on reflections, and eliminating reflections would also eliminate reverberation. However, single purpose theaters and opera houses, such as you saw in Figure 11.1, are usually built with boxes and balconies which hinder reflection, and some of these spaces have good acoustics.

Nodes and antinodes are the locations of intersecting maximum amplitude pressure fluctuations, either fully out of phase, adding to give locations of atmospheric pressure, or fully in phase, adding to produce maximum pressure fluctuations. If the time it takes a high to return to its source is an even number of half periods, the next high will be leaving the source just as a previous high's reflection returns there. This produces an antinode, and the source is said to be "in phase" with its reflection. If it takes an odd number of half periods, the reflected high will meet the source just as it is producing a low. This causes a node, and the source and reflected wave are "out of phase." In either case a standing wave results. Just what you don't want.

It might be possible to eliminate some of these resonances in a room by moving the sound source and thus changing the time it takes the wave

Chapter 11

to return. If the sound source is at the position of a node or antinode, moving it away from one of these will cause conflicting conditions: the sound source produces its maximum pressure fluctuations at a location where they do not exist for the room's resonances. This is another example of clamping out modes. If you know the particular resonance you want silenced, you can calculate the locations of its nodes and antinodes and not put the sound source there. Walls are always pressure antinodes, so if your sound source is a loud speaker try a location away from the wall. An orchestra on the other hand is a big extended sound source, often bigger than most of the wavelengths of the sound it is making. It is too big to be at the position of a room's possible node or antinode and experimenting by moving it from place to place will achieve minimal acoustic improvements not be worth the considerable irritation.

• QUESTION 11.6.   The previous paragraph said that there could be a maximum amplitude standing wave if the sound source is located at either a node or antinode of that standing wave.

At first glance this seems reasonable if the source is at an antinode, but why is this also true if the source is at a node? •

• QUESTION 11.7.   You have a rectangular room with dimensions 2.5 m, 7 m, and 10 m. Where would you place stereo loud speakers? To answer this question, first analyze the room to find its resonant frequencies and place the speakers to minimize their standing waves. Of course, the best way to eliminate room resonances is to listen with headphones. •

• QUESTION 11.8.   Repeat the above question for the same size room, which is open at one end of its longest length. •

Loudspeakers could also be used in rooms to combat standing waves caused by live music. This would require that the live music be recorded at its source, modified, and then be broadcast into the room from loudspeakers placed to fill in the nodes. This and similar possibilities will be discussed in the Section 11.2, "Active Acoustics."

11.1.2.2 Reverberation Time

Sound reflects from a room's boundaries and bounces back and forth within the room. This is reverberation. These boundaries, the walls or cavities in the walls, can be good reflectors so that the sound persists, or they can be poor ones, so that it dies sooner. The number of seconds it takes a pulse of sound to decay is its reverberation time (RT). Microphones and electronic instrumentation make exact measurements of the decaying sound, but the time it takes a hand clap to become inaudible is an often-used and easily-measured reverberation time even though different listeners might not agree exactly when the sound becomes inaudible.

During reverberation, sound bouncing off the room's boundaries may pass by the listener many times, each slightly less loud, depending on the fraction of the sound absorbed by the boundary and how much of it is dispersed by diffraction. What's left after absorption and dispersion will be reflected.

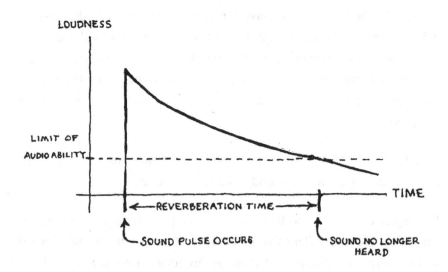

Chapter 11

The above graph shows how a typical listener hears a single short pulse of sound decay. Note, that the sound waves are still present after their loudness falls below the limit of audibility. Also realize that each listener will have his or her own limit of audibility, and therefore, his or her own perception of the music's reverberation time. This is another reason why listeners may not agree about what they hear. In this graph, and what follows, typical values will be used. The envelope, whose exact shape depends on the fraction of absorption and the amount of dispersion by diffraction, also determines the reverberation time.

A musician will play one note after another, controlling the loudness and duration of each. Reverberation time is not a measure of these things; it is only the room's contribution to the decaying loudness. Reverberation will continue the sound, causing tones to step on the attack of successive tones, generally adding warmth and reducing clarity.

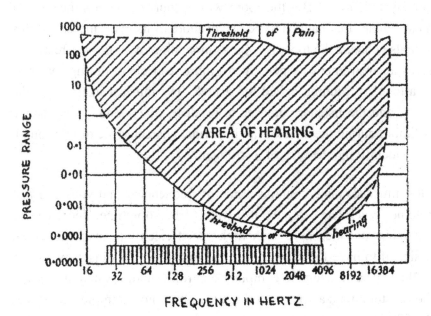

This graph shows the combinations of sound pressure fluctuations and frequency that a typical person can hear. Too much intensity and the sound will cause discomfort; too little and it will be inaudible.

Frequencies below about 50 Hz or above about 17 000 Hz won't be heard no matter what their intensity. Fix your attention on the lower curve showing the threshold of hearing. Note that it takes a lot more sound pressure to be able to hear very low and very high frequency tones. This indicates that, independent of how the sound is decaying, the listener's ear will tend to quit hearing high and low frequencies first, and give them shorter reverberation times. Here is a contribution to reverberation time that can vary from person to person. Also, if the sound consists of a range of frequencies, its timbre will change during the reverberation time.

You have already seen that during the short time in which a musical instrument establishes a tone many frequencies are present. Most of these are not modes of the tone the instrument has been adjusted to play, and they fail to establish a resonance in the instrument and quickly fade away. But the room will encourage the continuation of those pitches that happen to be one or more of its resonances, and this will also change the tone. It is the combination of all these effects that gives an instrument its perceived timbre, especially during that part of a tone's beginning called the **attack**. The musical instrument, the room, and the listener combine to produce "the sound." The change in timbre during the attack is an important characteristic of this sound, often identifying the type of instrument.

Most musical sound contains a mix of frequencies, and you have just read how each one can have its own reverberation time. These can be separated and measured with high speed electronic instrumentation designed to have the sensitivity of a typical ear. This instrumentation could be part of a general computer-controlled audio system that could then custom build a set of reverberation times. For example, increasing the *RT* for any frequency only requires that the computer generates that frequency and sends it to the loud speakers a little longer. Decreasing the speaker's loudness a little faster could shorten the *RT*. In either case the object is to change the envelope. Appendix B. "Computer

Chapter 11

Controlled Audio Electronics," describes such a system, and this and other audio modifications it could produce.

Some progress has been made in matching the qualitative descriptions of a room's acoustics with the room's reverberation times. Accepted recipes for changing a room's acoustics include[12]:

- to add warmth, increase the *RT* for the low frequencies. Or, if the sound is electronically amplified and you have loudness to spare, the *RT* for the higher frequencies could be reduced.

- to add liveliness, increase the *RT* for frequencies of about 200 Hz. The wavelength for 200 Hz is 1.8m and so you will need good reflecting obstacles much bigger than this. Or you could reduce the *RT* for 600 Hz waves. This will increase the room's liveliness by adding clarity to these higher frequencies. The wavelength of 600 Hz sound waves is about 0.6m, and obstacles about this size would diffuse them by diffraction.

- increasing *RT* will lessen the dryness of a room.

- decreasing *RT* will increase the clarity.

Appendix A describes and explains what happens to sound when it encounters obstacles, and indicates how reverberation time changes might be accomplished without electronics by controlling reflection. Perhaps a combination of changeable obstacles and computer-controlled audio will be necessary. There is a place where any reverberation time must be added if it is wanted. The outdoors, with no reflecting walls has zero reverberation. It is an "acoustically neutral room," and does nothing to the sound, which passes by the listener only once, directly from its source, giving it great clarity, extreme dryness, and practically no warmth. A room with perfectly absorbing walls is one, too. It is generally accepted that zero *RT* causes bad acoustics.

Part 3: Acoustics

11.1.2.3 Delay Time

Delay Time is the difference in the number of seconds it takes the sound to reach the listener directly from its source and by its other possible paths. Unlike reverberation where the same sound persists, delay time describes the sound arriving at the listener at separate times. Reverberation times can be as long as 3 or 4 seconds, but delay times are usually less than 0.5 second. Both reverberation times and delay times that are this long produce bad acoustics. Figure 11.6 shows an example of several possible sound paths in an auditorium.

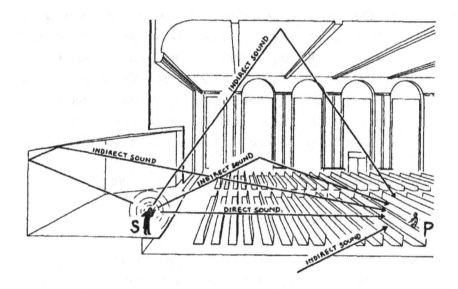

Figure 11.6 Sound Paths in an Auditorium

The listener at point P hears the sound first directly from its source, S, which is the shortest path, and later when the first reflected sound arrives. The **direct sound's** path and several indirect sound paths are shown. This listener receives the sound at four different times and from different directions. How does this change the listener's perception of the sound?

Chapter 11

If the sounds all arrive within 0.035s after the first, the listener will hear them as one, coming from the direction of the first to arrive. The direct sound should arrive within this interval, and better yet, be the first to arrive.

If the sounds all arrive between 0.035s and 0.060s after the first, the listener will hear them as one but will notice that they are arriving from the different directions. Listeners say that they prefer this type of stereo environment. All the indirect sound should arrive within this interval.

If the delay time is greater than 0.060s after the first, the listener will hear the sounds separately, an echo. Echoes are acoustical disasters.

These numbers are approximate and will vary somewhat from person to person and with different frequencies of sound[12]. But they do show that the delay time is an important part of a room's acoustics.

Figure 11.7 shows these relationships in graphical form.

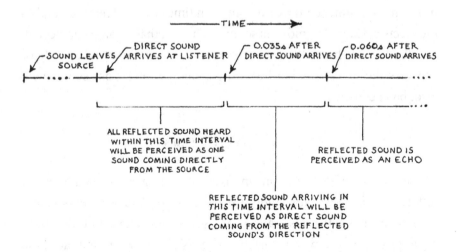

Figure 11.7 Delay Time and the Perception of Sound

Thus, you want to make all first reflections of a sound arrive between 0.035s and 0.060s after the direct sound. No first reflection should reach the listener after 0.060s. Can this be done? Probably so in small rooms, where reflections are not much longer than the direct path, but what about big rooms? There the path differences might be big enough to cause delay times greater than 0.060s, and it is also probable that listeners throughout the hall will experience different delay times. Perhaps quite different, and the worst seat could be front row center. It's a short distance to the musicians, and you might get a good view, but the reflections might reach you after a long journey. A seat near the rear might be better[13]. There the direct distance to the sound source will be long and thus closer to the length of any reflected path. On the other hand, you might be so far from the source that the sound will not be loud enough. This was one of the problems that lead to the first use of electronic audio in auditoriums. The solution was to put microphones near the source and loudspeakers in the back of the hall.

One problem solved, but another created! Recall that electric signals going from the microphones to the loudspeakers travel much faster than sound through the air: about 300 000 000 m/s versus about 345 m/s, which is almost a factor of a million times faster. The time it takes the electrical signal to move is so much shorter, that it can be neglected, and the listener hears the first sound from the nearby speakers. He or she sees the musicians but hears the sound coming from somewhere else: bad acoustics.

11.2 Active Acoustics

By now you might have an ambivalent feeling about sound wave reflections. They are absolutely necessary in wind instruments where they help create the resonances of the played tones, but can ruin, or improve, a room's acoustics. A possible way around acoustic difficulties might be to make the room acoustically neutral, without any resonances or reflections. The sound would pass the listener only

Chapter 11

once, directly from its source. As stated above it would have clarity, but probably be described as "dry" or "lacking warmth." The sound may need some reverberation time, or even a set of reverberation times for different frequencies. However, except for differences in loudness because everyone is not the same distance from the source, an acoustically neutral room will give everyone in the audience the same characteristics of the music as far as reverberation times, resonances, and delay times. There will be none. This is not yet good acoustics, but is certainly is an improvement over an acoustics that varies from place to place within the room and depends on the type of music. An acoustic application of the Hippocratic Oath, "First, do no harm." suggests that an acoustically neutral room might be a good starting place from which satisfactory acoustics could be created.

Would it be possible to have a room in which the musicians themselves could specify the acoustics? Acoustics that could change from instant to instant if necessary: **active acoustics**.

A plan for obtaining this contains two almost independent parts: first, make your room acoustically neutral; and second, add the acoustical characteristics desired.

An acoustically neutral room might be thought of as a blank canvas on which the sound pictures will be painted. Canvas accepts most painting styles, and our room must accommodate varied musical presentations. This requires being able to give the room different types of acoustics, and in an acoustically neutral room this means that all the musical sound except the direct sound would be adjusted. Modern **computer-controlled audio electronics** can do this. Such manipulation of the sound is controversial, but it is the norm in contemporary recording studios.

Musicians and musical organizations can be recognized by their sound, and each may have its advocates and detractors. There is no agreement now, and probably never will be, about what constitutes best sound.

Part 3: Acoustics

However, audio and computer engineering have advanced to a level where it is possible to achieve a variety of acoustics according to the wishes of the musician or the listener, or both. What can be done here is to show how to understand, analyze, and modify sound, and leave it up to the musician or listener to specify what the sound should sound like. You sing in the shower because you sound good there even though the acoustics are bizarre.

A computer-controlled audio system has the potential both to reduce the acoustic deficiencies in a room and to help create the type of sound wanted. It does all this by modifying the resonances and near wall effects in a room, by adjusting the reverberation times, by changing the loudness, and by changing the delay time. Building a room that had no reflections would eliminate the problems caused by incorrect ones; they would be gone. But the need for possibly changeable reverberation times, loudness, or proper delay times would not be met either. Electronic modification of the sound does seem to offer solutions.

Our computer-controlled audio system would do a twofold job: one, reshape the music's waveform by controlling the loudness and the reflections, resonances, and reverberation times; and two, create a satisfactory delay time, making sure that the listener hears the direct sound first. Here is a possible scheme.

Microphones receive the sound near its source and convert it into a varying electrical signal. This is sent to a computer where it is modified according to previously supplied instructions, or as required by one or more listeners, human or mechanical, in the room. The corrections are sent to loud speakers in the room and mixed with the flawed sound to give improved acoustics.

High speed computers can do this. While the direct sound is traveling through the air from its source to the listeners, the computer has time to make its calculations and send them to the loud speakers. Figure 11.8

shows a block diagram of such a scheme. Appendix B describes the computer operations in more detail.

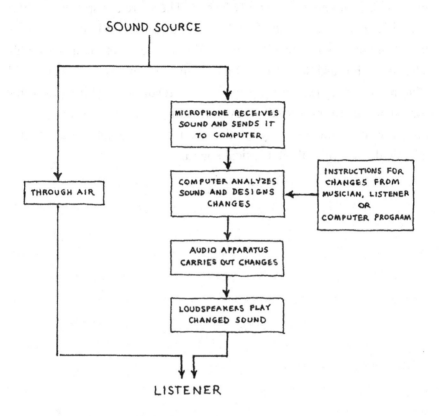

Figure 11.8 Block Diagram of Normal and Computer-changed Sound From Source to Listener

In smaller rooms the listeners are closer to the sound source, and the lack of sufficient loudness and incorrect delay times may not be problems. Big reflections, nevertheless, can cause resonances, near wall effects, and too long reverberation times. However, it's the rare hi-fi that does not have separate bass and treble adjustments (low and high frequency controls) and two movable speakers, and the listener usually can obtain satisfactory sound by adjusting and moving them.

Part 3: Acoustics

In larger rooms, such as auditoriums, lack of loudness and incorrect delay times or improper reverberation times can be important acoustical problems. Resonances, probably less so. These are the spaces that also contain the musicians, and often the major effort seems to be to improve the sound at its source rather than at the listener. Not always, though. The following article by Tom Manoff in the New York Times on 31 March 1991 describes the process and the results of making electronic modifications to the sound in several auditoriums. As you read it note the necessity and interplay of the changes to both the human and technical ways of making musical sound.

Chapter 11

Do Electronics Have a Place in the Concert Hall? Maybe

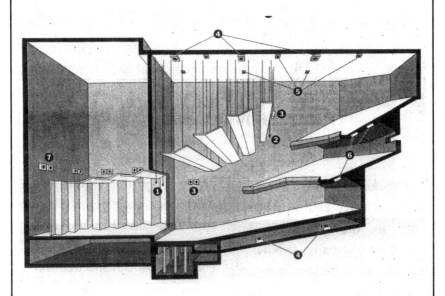

A Model of Electronic Architecture

This cross section of an imaginary concert hall, with the stage and orchestral shell at left, balconies at right and reflecting panels suspended from the ceiling, shows the elements of Jaffe Acoustics' Electronic Reflected Energy System.

❶ Stage Microphones: for balanced orchestral sound.

❷ House Microphone: to capture natural sound from the hall.

❸Early Field System: to provide the first reflected sound the ear hears.

❹Warmth Field System: to augment low-frequency sounds, adding richness and warmth.

> ❺Reverberation Field System: to lengthen reverberation time, making a "dry" hall more resonant.
>
> ❻Underbalcony System: to bring the natural sound of the hall to listeners under the overhang.
>
> ❼Stage System: to enhance what musicians hear back from the hall.

By Tom Manoff

EUGENE, Ore.

Architectural history may one day honor Silva Hall in the Hult Center for the Performing Arts in Eugene. Visually striking, it also affords first-class sound on most evenings. Not incidentally, the hall, completed in 1982, brings home perhaps the most controversial issue for the future of concert-hall design: "electronic architecture," combining traditional construction techniques and electronic "enhancement."

Because Eugene could afford only one hall to accommodate a variety of musical styles, the architectural firm Hardy Holzman Pfeiffer Associates of New York proposed a multipurpose acoustic design by Jaffee Acoustics of Norwalk, Conn. After a long debate as to whether there is any place in the concert hall for electronics, the original plan was approved. Silva Hall thus became the world's first electronically enhanced concert space conceived as such from the ground up.

Since opening night, the acoustics of Silva Hall have evoked both excitement and controversy. Rock, pop and Broadway musicals sound wonderful there. Opera also sounds fine. But for most symphonic concerts, the acoustics have proved unsuccessful.

When the talented young conductor Marin Alsop took over the Eugene Symphony in 1989, she requested a full demonstration of the Jaffe system and found it wanting. "I think it's a major problem," she said recently. She

Chapter 11

added that she found the enhanced sound lacking in "bottom," and worse, the orchestra sounded amplified. As a result, she turned the system off and began to experiment with the configuration of the orchestral shell.

By now, of course, the Eugene system is almost 10 years old, and its technology has grown antiquated. In the meantime, a second-generation Jaffe system used in the Alaska Center for the Performing Arts in Anchorage and a third-generation system in the Tennessee Performing Arts Center in Nashville have met with far greater success.

"An 'enhancement' system that failed a decade ago in Eugene has had greater success in Anchorage and Nashville."

Vital to the rapidly developing technology in Anchorage and Nashville is an internationally known sound technician named Gary Hardesty. As it happens, Mr. Hardesty lives in Eugene. When he saw the orchestra experimenting with Jaffe's shell, Mr. Hardesty offered his expertise. The orchestra, he said, politely ignored him. The management of the Eugene Symphony declined to discuss this and other issues connected with the hall.

■

With Silva Hall's electronics turned off, problems of another sort arose, since without enhancement the hall is, by design, relatively short of the reverberation that gives "warmth" to sound. Rock, pop and much theater, which use amplification, sound best in a very "dry," unreverberant, even "dead" hall. For these events, the Jaffe system is not used. Opera needs some reverberation, lasting 1.3 seconds, perhaps - enough for warmth but not so much that vocal clarity is lost. The Jaffe system in Sylva has variable settings; the one for opera is quite subtle and successful. But a symphony hall requires a reverberation time of from 1.8 to 2.2 seconds. Without it, orchestras sound dry, and symphonic music, which depends on a beautiful wash of sound, cannot succeed.

Ms. Alsop said she is willing to try anything. But she leans toward a purist approach. "To rely on a sound technician for your balance," she said, "is completely antithetical to the role of a conductor."

Her attitude is apparently not shared by the German conductor Helmuth Rilling, who comes to Eugene each summer to direct the Oregon Bach Festival. Mr. Rilling works closely with Mr. Hardesty, and together they occasionally coax a decent sound from the hall.

The Jaffe System, called ERES (Electronic Reflected Energy System), is the brainchild of Christopher Jaffe, who, as the president of Jaffe Acoustics, has been the main proponent of electronic architecture in the United States.

His work is well exemplified at Ravinia, the summer home of the Chicago Symphony, in Highland Park, Ill, where he designed Bennett Hall along traditional lines as well as the electronically enhanced Ravinia Festival Pavilion. Mr. Jaffe also included electronic enhancement in the restoration of the Circle Theater in Indianapolis, whose acoustics Harold Schonberg of The Times described as admirable

"I don't come to a hall with a preconceived notion," Mr. Jaffe said recently. "I offer the options of either traditional or electronic design."

■

The Evangeline Atwood Concert Hall in the Alaska Center for the Performing Arts was completed in 1988 by the same team that built Eugene's hall, and designed along similar lines. One look at the ceiling's reflective plaster panels (a distinct contrast with Silva's sound absorptive "basket" design) on a recent visit indicated that the sound, even without electronic enhancement, would be far livelier than Silva's. The first chords of Mozart's Overture to "The Magic Flute" from the Anchorage Symphony sounded round, rich and true. Although a longer reverberation time might be preferable, the natural sound, blended with the newer Jaffe technology, made for a beautiful orchestral acoustic.

Unlike Eugene's system, Anchorage's has no variable settings. It is either on or off. The involvement of a sound

Chapter 11

technician in orchestral balance is not an issue. Stephen Stein, the orchestra's conductor, said he was delighted with the Jaffe sound: "For a community our size, we feel no less than blessed."

Later in the week, the Anchorage Concert Association presented the pianist Garrick Ohlsson. Ira Perman, the association's executive director, said he considered the hall's natural acoustic fine, and didn't generally use ERES. But with the consent of the pianist, Mr. Perman agreed to an experiment. Mr. Ohlsson played the first half of his all-Chopin program with the system off, the second half with it on. In this test, heard from a "worst-case" seat, far left, under the first balcony and directly beneath one of the hidden speakers, the system was not entirely successful, seeming to vindicate Mr. Perman's decision to leave it off for solo performances.

Mr. Ohlsson began with the B-minor Scherzo. From the first notes, one knew the evening would be memorable. But by the time the pianist played the C-sharp-minor Nocturne (Op.27, No. 1), acoustical problems were apparent. The lower register, from middle C down, was warm, reverberant and perfectly responsive to the pianist's touch. The high octaves were brittle but acceptable. Least audible were the two octaves above middle C, no small problem in Chopin.

The second half of the program consisted of Chopin's 24 Preludes. With the system on, the troublesome octaves became more audible. But unexpectedly, the beauty of the lower registers disappeared into what could only be called an artificial sound.

■

Atwood Hall is a splendid auditorium for symphonic music, but an upgrade to the latest technology might some day be advisable, if the new sound of Andrew Jackson Hall at the Tennessee Performing Arts Center is any indication of the possibilities. Soon after the hall was built in 1981, the Nashville Symphony and its conductor Kenneth Schermerhorn, became unhappy with its dry and

Part 3: Acoustics

unresponsive sound. In 1990, Mr. Jaffe was called in to renovate it. The Nashville system, designed by Paul Scarbrough, a young architect on Mr. Jaffe's staff, was installed in less than three months for under $300,000.

When it was first demonstrated, the players reportedly broke into applause. "Just imagine if we were stuck with what we had," said Mary Vanosdale, the concertmaster, in a recent interview. "You can't imagine what it's like to be giving your all on stage -- and for what?"

To purists who criticize Mr. Jaffe's use of electronic enhancement, Mr. Schermerhorn said, "Well, more power to you if you have the unlimited funds to tear down this hall and build a new one."

In a concert performance of Mozart's "Abduction From the Seraglio," the orchestral sound from mid-parquet was excellent, rich, clear, not in the least artificial. The system dealt honestly with the soloists. Kenneth Cox, a world-class Osmin, sounded rich and powerful, although the acoustics were less kind to the tenor who sang Pedrillo. In the parquet, where the Jaffe system was most active, the acoustics revealed his inadequacies of style and manner and poor vocal quality. From the highest balcony, he could scarcely be heard.

■

Despite minor problems that will be corrected when adjustments are complete, the Nashville system seemed to represent not merely a local success but a significant new technology for the world. One reason for the improved quality was a higher digital sampling rate.

The sophistication of the Nashville system can best be explained in comparison to a CD player. Put simply CD technology measures sound 44,100 times per second. This number is called the sampling rate. In digital technology, the higher the sampling rate, the better the sound Those critical of CD sound have suggested that this sampling rate is too low. Higher rates are now being discussed.

Chapter 11

The first-generation system in Eugene doesn't even use digital sampling. The second-generation system in Anchorage samples the sound 68,000 times a second. And the up-to-date technology in Nashville includes a device that samples 3 million times a second. That device was invented and built by Mr. Hardesty.

The future of the Jaffee system seems clear. In a world of scarce resources, this technology offers an economical alternative to traditional modes of construction. But what about Eugene? Are its citizens, having paid for the prototype of a successful technology, stuck with a magnificent but flawed monument to the science of acoustics?

Some people question whether the Eugene hall should be fixed at all. Except for its symphonic sound, it is a success. Eugenians have a strong populist bent, and a city government hard pressed to maintain basic services has little extra money for the arts. Any expenditure that didn't come from private contributions might be criticized as elitist. And even if money were available, the question would remain how to fix the hall. Structural renovation would be expensive, but a third-generation Jaffe system could be installed for under $200,000.

■

Yet a far greater opportunity is available to Eugene, in the person of Mr. Hardesty. Although he has invented some of the world's most advanced electronics, he has never held an official position at the Eugene hall. His status is "stagehand." He has spent hundreds of hours, often unpaid, working to improve the hall. Asked recently why he bothers, he said simply, "It's my home. I care about the hall."

Edwin R. Smith, the city official who was in charge of building and for a time operating the Hult Center, has only praise for Mr. Hardesty. When asked why Mr. Hardesty has never been recognized officially, Mr. Smith replied, "That's a good question."

◻

The Jaffe System

The Jaffe Electronic Reflected Energy System (ERES) bears virtually no relation to traditional amplification, in which an audio signal (a voice or instrument, for example) is projected into a hall at a volume louder than the original source. In ERES, the primary and loudest sound the listener hears is the direct, natural sound from the stage, which mingles with very low of levels of electronically manipulated sounds from hidden speakers. The result re-creates the acoustic characteristics of a traditional concert hall.

The sophistication of the latest ERES technology, which samples and reproduces sound three million times a second, brings to mind certain philosophical quandaries: How many dots, for example, must be drawn and connected by lines to achieve a perfect circle? In a comparable way, the quality of this sampling system tends to break down the esthetic oppositions between "natural" and "electronic," "acoustic" and "artificial" and "real" and "reproduced."

A similar and related quandary: how many speakers would have to be mounted to re-create an infinite number of sound-reflecting points on the wood and plaster surfaces of a traditional hall? Christopher Jaffe has invented a means of processing the signals so a speaker that would normally be heard as a single source of sound flattens out into what the ear perceives as a section of wall. Theoretically, a complete Jaffe installation, with its multitude of speakers and interrelated systems, can be adjusted to re-create the acoustic response or of any architectural design. —T.M.

> Tom Manoff is a music critic for "All Things Considered" on National Public Radio.

Chapter 11

Manoff's article shows a possible auditorium equipped with microphones and speakers, but doesn't say much about what changes are made to the sound while it is in its electronic format between microphones and speakers. Nevertheless you can recognize uses for each set of microphones and speakers.

The Stage Microphones sample the sound as it leaves the orchestra. This is the sound the conductor and the musicians want the audience to hear. If microphones in the hall receive a different sound, corrections can be made. The sound from the Stage Mikes can also be sent to the Balcony Speakers if more loudness is needed there.

The House Microphones receive the sound at various places in the audience. This can be played back through the Stage System speakers to let the conductor and musicians hear what the audience is actually hearing. Corrections can be made, perhaps with a division of labor: humans in charge of the musicality, and the computer detecting and correcting obvious acoustical problems in the room. The necessary modifications to delay time, frequency balance, and reverberation time are played through the Early Field System speakers, the Warmth Field System speakers, and the Reverberation Field System speakers.

APPENDICES

APPENDICES

Appendix A.
Sound Becomes Less Loud: A Closer Look at the Structure of a Gas

Sound becomes softer as its high pressure regions become less high and its low pressure ones less low. When they both become atmospheric pressure the sound is gone. Sometimes this happens in air alone. Or the sound wave hits an obstacle and processes called **diffraction** and **absorption** cause the pressure changes. In order to understand how this happens you will have to know more about gases and pressure.

A.1 A Model of a Gas

You've already read that high pressure (meaning higher than atmospheric pressure; the words "high and low" describe pressures above and below atmospheric pressure) happens when the air molecules are packed together closer than normal. You will see that pressure changes can depend on the air's temperature changes too, but for musical sound this is not important. "Low pressure" is a synonym for "partial vacuum," and a region of air molecules spaced farther apart than normal is below atmospheric pressure and therefore a partial vacuum. A "vacuum" would have no gas molecules present.

Experience tells you that a lot of air can be added to a tire without changing its volume much even though this does increase its pressure. Add molecules and the pressure increases; take them away and it decreases. You can't do this to a solid. It's almost impossible to double the amount of material in most solids without doubling their volumes.

Appendices

Yet everything is made from molecules; solid ice and the gas called water vapor are made from the same kind. But, a cubic meter of ice weighs a lot more than a cubic meter of water vapor. This is strong evidence that there are a lot more molecules in the ice, and that they must be a lot closer together. There is lots of room left in a gas for more molecules.

DEMONSTRATION VI, in Chapter 1, used a closed syringe to show the elasticity of air. The plunger was pushed down compressing the air in the tube. When it was released the compressed air pushed it back up again. Now you know the term "compressed air" means a greater than normal concentration of molecules. More molecules hit the plunger each second from inside than from outside, and this caused the plunger to rise. If you pull up on the plunger, increasing the volume of the same amount of air in the tubing, the concentration of molecules there will decrease and now there will be more collisions from the outside air. The plunger will be pushed back down. The explanation of pressure caused by the motion of molecules describes both cases. The plunger in the demonstration was real, but it could have been made less and less massive until it was no longer there except as the boundary between regions of different pressures. This is what's happening within sound waves.

It took most of the nineteenth century to discover that matter is made of molecules, and that the distance between them, in the air around us, is about 34 times the size of an air molecule itself. Indeed, most of the volume of air is empty space.

It would not be correct, however, to call these individual pieces of empty space vacuum. Vacuum is one of those **macroscopic** words, and so is pressure, as you will see, that can be applied only to large collections of things. The space between two air molecules is not a vacuum because it is not big enough.

How can these molecules, so far apart, produce pressure?

Appendix A

DEMONSTRATION XXIV: Mechanical Model of a Gas

Apparatus: Kinetic Theory of Gas Apparatus

Here is a mechanical model of a gas.

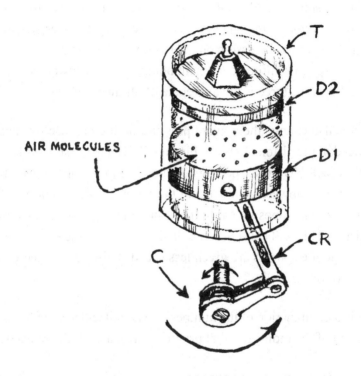

A clear plastic tube, T, contains two close fitting, but not snug discs, D1 and D2. The small hard spheres between D1 and D2 represent air molecules. The volume between the discs is the volume of our air and the spheres are the air molecules. Disc D1 is a piston attached to a rotating crankshaft, C, by a connecting rod, CR. The crankshaft rotates and piston D1 moves up and down. Its rate and speed can be adjusted by changing the rate of rotation of the crankshaft.

- 213 -

Appendices

The spheres bounce around as the piston moves up and down. They hit disc D2 and raise it. You can adjust conditions by changing the crankshaft's rotation rate, by putting additional weight on top of disc D2, or by adding or removing spheres. Various combinations of these will change the volume between the discs. Our model of a gas is this mostly empty volume and whatever spheres are in it.

Add more spheres and it takes more weight on the top disc to keep the same volume. More spheres need more weight, fewer spheres less weight. The upward pressure on the bottom of the top disc comes from the flying spheres hitting it. Each hit bumps the disc upward a little, and the total upward force is the aggregate of all these bumps.

You can also change the upward pressure on the top disc by changing the speed of the piston. A slower piston will whack the spheres less hard and they will have less speed when they bump the upper disc. It will not be driven upward as much and the volume will decrease. A faster piston does the opposite. Or, you can put a layer of softer material on the piston. This will cause softer collisions between the spheres and the piston, and the spheres are given less speed. This too will decrease the volume of our gas.

So, it is a combination of the number of gas molecules in a volume and the speed of the molecules that produces pressure. Both are needed.

In any case, a constant pressure needs many and continuous hits by the spheres. Too few spheres make the time between hits so long that the top disc would be jumpy, and the word "pressure" denoting a single continuous quantity would no longer apply to our gas. This is why I said that pressure is an example of a **macroscopic** property of a thing, one that needs quite a lot of that thing present before it can be said to exist. The vacuum described previously is also a macroscopic property. The alternative is called a **microscopic** property. It is there in the smallest subdivision of a thing, or so the story goes. A wood molecule is not a

Appendix A

tree; a proton or electron is not a molecule; and a quark is not a proton. Have we yet discovered the microscopic properties of a tree?

A.2 The Ideal Gas Law

Hard spheres, far apart, is a satisfactory model of a gas. It does not include, nor does it need, the complex structure of a real air molecule. No atoms with their clouds of electrons around a tiny nucleus, our air molecules are simple hard spheres. Certainly the fine details of the behavior of a gas can be explained more correctly using a model that includes the detailed structure of the molecules, but not in this book. Our hard sphere model works well for air as it does its everyday musical sound operations.

The above demonstration cannot show the exact relationships between changes in the number of molecules and changes in the pressure, nor exactly how the pressure changes as the molecules' speed changes. These can be obtained from careful experimentation with much better apparatus. Such experiments show:

1. changing the number of gas molecules in a fixed volume by some percentage will change the gas' pressure in that volume by the same percentage. You must keep the gas' temperature constant during this.

2. changing the average value of the squared speed of a fixed number of molecules in a fixed volume by some percentage will change the pressure by the same percentage. If you are dubious about this being a directly experimental result you should be. There is no direct way, say with a meter stick and a stopwatch, to measure a gas molecule's speed.

3. pressure, p, volume, V, number of molecules, N, and temperature, T, of a gas are related by a formula called The Ideal Gas Law:

Appendices

$$pV = kNT. \qquad (A.1)$$

Everything in this formula should be expressed in metric system units. For example, the volume, V, must have the units of cubic meters, m^3. In addition, the pressure, p, and the temperature, T, use absolute scale. Pressures are measured from vacuum and temperatures from absolute zero. The metric units for pressure and temperature are rather ungainly and not in common use. So, let's defer mentioning them until they are really needed.

The symbol k represents a constant, just like the number of cents in a Euro, 100 cents/Euro, and is needed in a similar way to keep the amounts correct.

The Ideal Gas Law was developed from experimental work done in the seventeenth and eighteenth centuries, long before people knew that air, or any gas, was made of molecules. However, it does contain N, the number of molecules, and so it is a more modern nineteenth century statement.

• QUESTION A.1.   How do you suppose it is possible to do the experiment referred to in the first experimental result? In particular, how could you change the number of molecules by a fixed percentage? This isn't a job to be done with tweezers. You can assume that it's easy to measure the pressure, temperature, and volume; these are macroscopic quantities. But, there is no way to count the number of molecules. •

• QUESTION A.2.   In Chapter 1 it was claimed that packing more air molecules into the same volume would raise the pressure of the air in that volume. DEMONSTRATION VI showed that decreasing the volume of air containing a fixed number of air molecules also would increase the pressure of that air. In both cases the number of air molecules per air volume is increased. Argue that the Ideal Gas Law,

$$pV = kNT$$

Appendix A

predicts both of these pressure increases.

Hint: In the first case the volume and the temperature stay constant. N changes. Doing the algebra to isolate p on the left side of the equal sign produces,

$$p = (kT/V)N.$$

Everything in the parenthesis is constant; so changing N must change p. But, by how much?

In the case of the syringe V changes while N and T stay the same. This means that the terms on the right hand side of the equal sign in the Ideal Gas Law don't change, i.e.,

$$pV = \text{constant.}$$

What happens to p if now V is made smaller? In particular, if V becomes half as big, by how much will p change? •

A.3 The Kinetic Theory of Gases

An accepted theory states that a gas' temperature is proportional to the average value of the squared speeds of its molecules. The second experimental result above is a combination of this theory and the experimentally based Ideal Gas Law and thus, not wholly experimental.

The average value of the squared speeds of the molecules is represented by the symbol $(v^2)_{ave}$. It could be calculated as follows: First, you would have to know the speed of every air molecule in whatever volume in which you are interested. Next, square the values of these speeds and add these together. Then, divide by the number of molecules in your volume. That's it; you have the average value of the squared speeds of your molecules. Don't try it though; it's an impossible task, not only

Appendices

because there is no way to measure the speed of each molecule; but also because there are billions and billions of them in even a small volume. Instead, $(v^2)_{ave}$ is obtained from the Kinetic Theory Equation, (A.2), shown below.

• QUESTION A.3. Combine the accepted theory that says that a gas' temperature is proportional to the average value of the squared speeds of its molecules, and The Ideal Gas Law, to produce a statement or equation that shows the above second experimental result.

Hint: The accepted theory written as a mathematical formula is

$$T = (\text{constant}) \, (v^2)_{ave},$$

where "constant" is the fixed value that converts the amount of the temperature, T, into the proper amount of the average value of the squared speed of the molecules, $(v^2)_{ave}$.

Next, note which terms in The Ideal Gas Law are kept unchanged in the statement of the second experimental result, i.e. which parts of the formula,

$$pV = kNT,$$

must not change for the second experimental result to apply. Must V stay constant? How about N? Yes, both must. So, rewrite The Ideal Gas Law with all the constant terms in one group between parentheses,

$$p = (Nk/V) \, T.$$

With this equation and the one above for T you should be able to produce the second experimental result. Do it. •

The theory connecting the molecules' speed and the gas' temperature is called the Kinetic Theory of Gases. It is one of the triumphs of science. Its assumption that a gas is a collection of molecules and not a dilute

Appendix A

continuous liquid opened the doors into modern physics. The Kinetic Theory of Gases predicts that

$$\frac{1}{2}m(v^2)_{ave} = \frac{3}{2}kT$$
(A.2)

where, m is the mass of a molecule, and $(v^2)_{ave}$ is the average value of the molecules' squared speeds. Here is an equation connecting two microscopic properties of a gas, the mass of one of its molecules and its speed, to a macroscopic property, the gas' temperature. This also makes the Kinetic Theory of Gases equation important: it states a relationship between the microscopic properties on the left side of the equal sign (the molecule's mass and its speed), which are impossible to directly measure, and the macroscopic temperature of the air on the right side, which is easily measured. Temperature is another of the macroscopic quantities. It makes no sense to talk about the temperature of a single molecule. A single molecule is just not a big enough collection. The Kinetic Theory of Gases equation is true only if there are enough molecules present so that a good average value of v^2 exists. A good average is another macroscopic property. The good average value for flipping a coin is half heads, half tails. Flip it only once and you will not get half a head.

Equation (A.2) also makes a startling prediction: if a gas is at absolute zero temperature, all molecular motion has ceased. Look at this equation again and see that if T is zero the part of the equation to the right of the equal sign is zero. Since this is an equation, the left hand side must be zero too. There are only two things there that could be zero: the mass of molecule or its average squared speed. Zero mass would mean no molecules, and no gas, so all that's left to be zero is the average squared speed. At absolute zero temperature gas molecules have zero speed: they are at rest. At any other temperature they will be moving. Their speeds increase with increasing temperature.

Appendices

The Ideal Gas Law states that when the gas' temperature is absolute zero, which makes the right side of The Ideal Gas Law zero, the gas' pressure will be zero too. This makes sense. Pressure is caused by molecules whacking something, and without any speed the molecules won't do any whacking. Our description of pressure caused by molecular motion has now taken on mathematical forms: The Ideal Gas Law, (A.1), and the Kinetic Theory of Gases equation, (A.2).

The metric absolute temperature scale is called the Kelvin scale after William Thomson, Lord Kelvin, (1824-1907). Each degree Kelvin, K, is the same size as a Celsius degree but the zeros are different. A table will show some temperatures expressed in Kelvin and Celsius degrees.

TEMPERATURE

K	°C	
373	100	WATER BOILS
273	0	WATER FREEZES
0	-273	ABSOLUTE ZERO. MOLECULAR MOTION STOPS

• QUESTION A.4.   Locate a biography of William Thomson, Lord Kelvin, and discuss why he has a temperature scale named after him. •

By now you might be wondering why the values of a sound's high and low pressures have not been stated. Also, just how big are the excesses and deficiencies of the molecular concentrations that cause these pressures? One reason is that the unit of pressure in the metric system is so ungainly that even scientists shy away from it. It's called the Pascal, Pa, Blaise Pascal, (1623-1662). Normal atmospheric pressure is 1.0×10^5 Pa. A bicycle tire gauge would need its scale marked from one hundred thousand to one million Pa! The pressure unit called the bar is often preferred. One bar is one normal atmospheric pressure, in other words 1 bar = 1.0×10^5 Pa, and your tire gauge would have a scale from one to ten bars. However, unless you are taking ratios of pressures,

Appendix A

absolute pressures in units of Pa must be used when you are calculating something using The Ideal Gas Law.

• QUESTION A.5.   Locate a biography of Blaise Pascal and discuss why he has had the metric unit of pressure named after him. •

Just like many microphones, most pressure gauges contain a thin diaphragm with one surface connected to the atmosphere and the other to the pressure to be measured. When these two pressures are different the diaphragm flexes and the amount of deflection, either steady or rapidly changing, is a measure of the unknown pressure or the sound. Such a device will indicate zero pressure when the unknown pressure is atmospheric pressure. This is called "gauge pressure." Partial vacuums are negative gauge pressures. The absolute pressure used in The Ideal Gas Law is measured from a vacuum and is always positive.

A Table will show some pressures expressed in different scales and units.

PRESSURE				
ABSOLUTE		GAUGE		
Pa	bar	Pa	bar	
8×10^5	8	7×10^5	7	BICYCLE TIRE
1×10^5	1	0	0	ATMOSPHERIC
0	0	-1×10^5	-1	VACUUM

Returning to the question of what are the pressures in a sound wave, it is possible to make a satisfactory estimate of them by analyzing the operation of the cavity-controlled oscillator in a particular instrument, the clarinet. The high pressure parts of the resonant sound wave in the bore push against the reed and open it periodically for the next puff of air to enter. How much pressure is needed? Begin answering this by estimating how many coins (or slices of prosciutto) would have to be hung from the reed to open it. It seems to me that about 100 g of coins (or about 6 slices) would surely do it. Next I looked at a clarinet mouthpiece and saw that the area of the reed in contact with the sound wave was about 10 cm². From these two bits of information I calculated the needed pressure in bars. The formula is

$$\text{needed pressure} = \frac{100 \text{ g} \left[\frac{1 \text{ kg}}{1000 \text{ g}}\right] (10 \text{ m/s}^2) \left[\frac{1 \text{ Pa}}{1 \text{ kg/m s}^2}\right] \left[\frac{1 \text{ bar}}{1.0 \times 10^5 \text{ Pa}}\right]}{(10 \text{ cm}^2) \left[\frac{1 \text{ m}}{100 \text{ cm}}\right]^2}$$

$$= 1 \times 10^{-2} \text{ bar}$$

$= 1 \times 10^{-2}$ bar.

All the quantities in square brackets convert units:

1 kg = 1000 g

1 m = 100 cm

Pa = 1 kg/m s²

1 bar = 1.0x10⁵ Pa

The 10 m/s² in the parenthesis is used to calculate the weight of the 100 g mass, after its units have been converted into kg. The 100 g and the 10 cm² are the only real data, everything else are conversion factors.

So, our answer is 0.01 bar, and the high and low pressure fluctuations in our sound wave are only about 1% of the normal atmospheric pressure.

Appendix A

How big a deviation from normal molecular concentrations will cause this? The Ideal Gas Law tells us that if you change the pressure so that the left hand side of the equal sign changes by 1%, so must the right hand side change by the same amount. Assuming that the temperature stays the same, the number of molecules, N, must increase by the needed 1%, and so both of these fluctuations are also only 1% from normal. For every 100 air molecules only one is added or taken away. The sound wave did not change anything much at all, and all the gas' properties remain almost the same whether or not musical sound is present.

Now that you've seen The Ideal Gas Law and the speed-temperature Kinetic Theory of Gases equation it would be a waste not to list some of the properties of air that can be calculated from them once you know the values of the constant k, and the measured density of air at usual atmospheric pressure, and temperature (293K or 20°C).

$$k = 1.4 \times 10^{-23} \frac{\text{kg m}^2}{\text{s}^2 \text{ K}}$$

density of air @ 20°C and normal pressure = 1.2 kg/m³

Now you have everything you need to find that:

mass of an air molecule = 5.0×10^{-26} kg

average speed*, $\sqrt{(v^2)_{ave}}$, of an air molecule @ 20°C≈500 m/s**

---

* Calling this the average speed is not quite correct. Its proper name is "the root mean square of the molecules' speeds," abbreviated "the rms speed of the molecules." It is the square root of the previously described average value of the squared speeds of the molecules. You can see that trying to make the names accurately describe the speed is leading to ridiculously convoluted phrases, and so "average speed," which is not very wrong, is used.

** The wiggly equal sign means "is about equal to." Sometimes an approximate value shows a relationship just as well, and is easier to read.

Appendices

number of molecules in a cubic meter @ 20°C
and normal pressure = 2.4×10^{25} molecules/m^3

average distance between molecules @ 20°C
and normal pressure = 3.4×10^{-9} m

approximate number of collisions per second a
single molecule makes with the others = 1.5×10^{11}/s.

Adding one more piece of information, the approximate size of a molecule, allows you to find the average distance between air molecules at 20°C and normal pressure:

approximate diameter of a molecule = 1×10^{-10} m.

The approximate distance between molecules is about 34 times the molecule's size. They are close together on a size scale for everyday things, but far apart relative to their size. It seems that everything about the sub-microscopic world of a gas is described using numbers so large or small that they are meaningless. Everything, that is, except the average speed; 500 m/s is not much larger than the speed of sound, as it must be if our description of sound propagating by collisions between molecules is correct. Perhaps you were surprised that air is so dense. That is the normal air near us. As altitude increases and the earth's gravitational pull on it lessens, its density and pressure decrease to the zero values for the vacuum of space.

The details of most of the calculations of these results have not been shown, but everything needed has been given. You can test your understanding of The Ideal Gas Law and the Kinetic Theory of Gases equation by working out the answers to the following questions. Feel free to use words, sketches, and formulas doing this; it's a standard scientific way of presenting information.

Appendix A

• QUESTION A.6.   You add air to a tire and its pressure increases. What's happening? •

• QUESTION A.7.   Use The Ideal Gas Law, equation (A.2), and the given values for k, the molar volume, and the density of air to calculate,

-the mass of an air molecule

-the average speed of an air molecule @ 20°C

-the number of molecules in a cubic meter of gas @ 20°C and normal pressure

-the average distance between gas molecules @ 20°C and normal pressure

-the approximate number of collisions per second that a single molecule makes with the others. •

• QUESTION A.8.   Show that The Ideal Gas Law and the Kinetic Theory equation can be combined to show that "changing the average value of the squared speed of a fixed number of molecules in a fixed volume by some percentage will change the pressure by the same percentage." •

The purpose of this appendix, so far, has been to introduce and make familiar the molecular structure of a gas: hard spheres fairly far apart and moving randomly as they collide with each other. The sources of sound create local excesses and deficiencies in the concentrations of the molecules, and the motion of these concentrations away from the source is the sound wave. This is the way musical instruments create sound. The excess molecules, just because there are more of them, beat more often against the nearby normal molecular concentrations causing them to be driven closer together. This relieves the pressure in one place and produces excess pressure nearby, and this is the way high pressure

Appendices

regions move. Just the opposite happens as the low pressure regions move. The individual molecules travel hardly more than 34 diameters before they hit others and stop or are deflected. They don't move very far at all. It's the regions of abnormal concentration that travel the long distances from the sound source to the listener.

The general process of sound becoming less loud is now understandable: something makes the traveling high and low molecular concentrations become more normal atmospheric concentrations. Let's now take a look at what that something, or somethings, can be.

A.4 Sound Changes Loudness

A.4.1 Sound in Air Alone

Sound can become less loud while traveling through air alone. The sound wave expands as it leaves its source and both the excess and deficient concentrations are spread over bigger wave fronts, thus dissipating them.

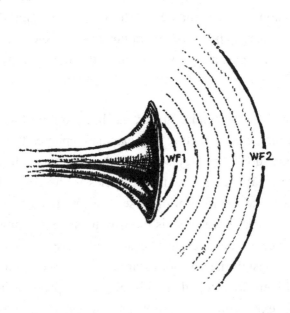

Appendix A

This sketch shows a high pressure wave front spreading as it leaves the horn. The abnormal concentration now at wave front WF1 will be later distributed over wave front WF2. If the sound source is small, or if you are far enough away from it so that it looks small, the sound wave's expansion can be visualized as being the increasing surface of a round balloon being inflated. The same amount of rubber has to make itself into a bigger surface. In our case the excess air molecules are the rubber; and just as the rubber surface must become thinner and thinner, the excess molecules get farther apart until they have the normal concentration of atmospheric pressure. When you recall that the formula for the surface of a sphere having radius r is $4\pi r^2$, you can understand that the abnormal molecular concentrations change as $1/r^2$. This relationship is often called the "one over r squared law" or the "inverse square law," but you can see that it's just a bit of geometry.

• QUESTION A.9.   Make the argument that "the abnormal molecular concentrations change as $1/r^2$." •

A.4.2 Sound Hits Obstacles--Reflection, Diffraction, Absorption

Sooner or later a sound will hit some obstacle, and other kinds of loudness loss can occur. These can be put into three categories: by reflection, by diffraction, and by absorption.

A.4.2.1 Reflection

You've already read quite a lot about reflection; let's continue with it. Certainly sound waves reflecting from a wall will cause standing waves or near wall effects, but in these, for every antinode there will be a nearby node. All the sound is still there, just separated now into regions of more and less loudness. So, this is not a general lessening of the loudness. However, if the reflection directs some of the sound into an "unoccupied" volume, the general loudness there will increase. Now there will be less loudness in the volume where the sound would have arrived prior to the reflection. This is a way to increase, or decrease,

Appendices

loudness; but it, too, is a zero-sum procedure. The sketch below shows reflectors shaped to redirect the sound this way.

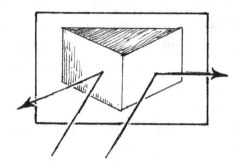

A.4.2.2 Diffraction

Diffraction is the name of the process in which wave fronts are bent and transformed into new shapes. The details are quite complicated, but the effect depends largely on whether the obstacle is bigger or smaller than the wavelength. Luckily, diffraction can happen to any wave, and by picking one you can see, a water wave, a visual demonstration can show some of the details.

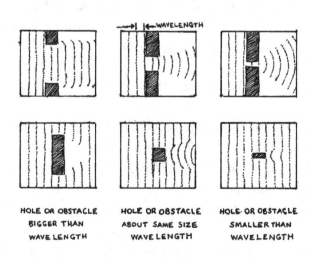

Figure A.1 Diffraction of Water Waves

Appendix A

Figure A.1 shows water waves diffracting from holes and obstacles. A wave coming from the left is broken up by an obstacle or a hole, and then recombined by a complicated superposition into a new shape (but not a new wavelength or frequency). Small objects distort the wave's shape more, but influence only a smaller portion of the wave front, and diluting a little bit of the wave into a larger volume does not change the loudness much. Big objects produce shadows, places where there is none of the wave. When the wavelength and the obstacle or hole are about the same size, diffraction causes effects between these two extremes, breaking up the wave and making it more irregular. This will hinder a resonance or a near wall effect. Just what we might want; but there are limits. Look at the range of sound wavelengths in Table 11.2 and consider the sizes of obstacles needed. Nevertheless, many of the screens and baffles you see in large rooms are there to diffract the sound from otherwise reflecting walls. Boxes, balconies, and even the audience in their seats can do this too. The boxes are holes in the walls; the balconies and the people are obstacles.

Light has an extremely short wavelength, and all normal objects are huge in comparison. Thus we expect, and get, shadows; and cannot see around corners. Sound, with its everyday sized wavelengths, diffracts from most things around us; and this helps us hear around corners. Reflection, if present, also helps.

• QUESTION A.10.   What frequency sound will be most strongly diffracted by the people in an audience? •

A.4.2.3 Absorption

In addition to reflection and diffraction, there is the possibility of the sound wave hitting the obstacle and the air molecules losing some of their speed there by making so called "soft collisions." And, less speed means less pressure. This is absorption, and is another way to lessen a gas' pressure.

Appendices

The exact interaction between the sound wave and the absorber depends on the details of the absorber's construction. Sometimes this is explained by describing collisions between the individual molecules in the sound wave and in the absorber. Sometimes the additional intermolecular forces in the solid absorber cause each air molecule to interact with larger parts of the absorber.

An analysis of a macroscopic demonstration of hard and soft collisions will show a little more detail.

DEMONSTRATION XXV: Hard and Soft Collisions

Apparatus: Hard Wall, Soft Wall, Rubber Ball

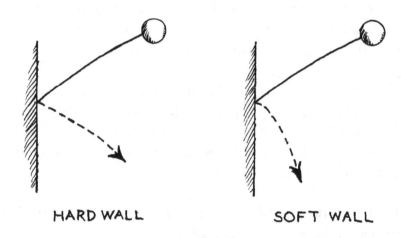

HARD WALL   SOFT WALL

The hard wall is plastered brick. The soft one is covered with a layer of expanded plastic or sponge. The ball bounces back from the wall in both cases, but not with the same vigor. During the collision both the walls compress enough to cause a force to stop the ball, but the hard wall does not compress as far. It applies the stopping force over a shorter distance and more quickly. This takes a bigger stopping force. The soft wall allows the ball to be in contact a longer time, and this in

Appendix A

turn gives this wall more time for other available methods of energy exchange to occur. Some of the ball's motion was changed into the soft wall's quite different motion. The whole wall moved with a different frequency. The soft wall will not return all the original energy to the ball and the rebounded ball's motion is decreased. You see that it had less speed, and therefore less ability to compress as much in subsequent collisions. This is analogous to a sound wave having less large pressure fluctuations after an absorptive collision. The sound is not as loud.

In this demonstration the ball represents a high pressure region of the sound wave. You may well ask, "What is the mechanism for eliminating the lows?" Remember that objects can expand as well as compress. A low pressure "ball" hitting the wall would cause the wall to expand. Again, some of the ball's energy of motion would be transferred to the wall. In both of these collisions there will be less energy available to the rebounding ball.

Collisions between sound waves and absorbers have this same general character. Some of the wave's energy is changed into some other form and the reflected part of the sound is less loud.

Perhaps you have heated a strip of metal by repeatedly flexing it. Some of the energy you supplied to make the metal move has changed form into heat. This is a one-way process. Heating the metal will not cause it to begin flexing. And playing noise into an absorber will not create musical sound.

A rule of nature called The Second Law of Thermodynamics states that energy by itself will not organize itself into a more ordered form. In the above description a steady heating will not cyclically bend metal; and noise, by itself, will not organize into musical sound. The First Law of Thermodynamics says that although energy can change form, the total amount remains the same. As their names imply, both of these laws were first used to explain processes involving heat. They are now

Appendices

considered to be quite universal and fundamental laws of nature. The First Law is also given another name: The Conservation of Energy.

Surface Treatment	Absorptivity at Frequency					
	125	250	500	1000	2000	4000
Acoustic tile, rigidly mounted	.2	.4	.7	.8	.6	.4
Acoustic tile, suspended in frames	.5	.7	.6	.7	.7	.5
Acoustical plaster	.1	.2	.5	.6	.7	.7
Ordinary plaster, on lath	.2	.15	.1	.05	.04	.05
Gypsum wallboard, ½" on studs	.3	.1	.05	.04	.07	.1
Plywood sheet, ¼" on studs	.6	.3	.1	.1	.1	.1
Concrete block, unpainted	.4	.4	.3	.3	.4	.3
Concrete block, painted	.1	.05	.06	.07	.1	.1
Concrete, poured	.01	.01	.02	.02	.02	.03
Brick	.03	.03	.03	.04	.05	.07
Vinyl tile, on concrete	.02	.03	.03	.03	.03	.02
Heavy carpet, on concrete	.02	.06	.15	.4	.6	.6
Heavy carpet, on felt backing	.1	.3	.4	.5	.6	.7
Platform floor, wooden	.4	.3	.2	.2	.15	.1
Ordinary window glass	.3	.2	.2	.1	.07	.04
Heavy plate glass	.2	.06	.04	.03	.02	.02
Draperies, medium velour	.07	.3	.5	.7	.7	.6
Upholstered seating, unoccupied	.2	.4	.6	.7	.6	.6
Upholstered seating, occupied	.4	.6	.8	.9	.9	.9
Wood/metal seating, unoccupied	.02	.03	.03	.06	.06	.05
Wooden pews, occupied	.4	.4	.7	.7	.8	.7

Table A.1 Absorption Coefficients of Various Surfaces and Objects

Table A.1 lists the absorptive properties of various surfaces and objects for several frequency sounds. The values of the absorptivity are the fractions of the object's total area that can be considered to be a perfect absorber. These values are approximate, but do give you an idea of what materials or objects are good, or bad, absorbers.

A horn's bell might be called a soft obstacle. The gradual expansion of the tubing constrains the sound wave to expand rather slowly, hindering any sudden pressure changes that could cause intense reflections. Would it be correct to say that the bell is, from the standpoint of the sound in the horn, a good absorber? No, there is no change in the type of energy there. It's the same frequency sound wave inside and out.

Appendix A

• QUESTION A.11.   You now have a list of general names for objects or materials that can change a sound's loudness: reflector, diffractor, absorber. Should the name "transmitter" be added? •

A.4.2.4 Impedance

Collisions with partial reflection-partial transmission at a boundary are not limited to sound. Other types of waves and pulses, such as light, electric, hydraulic, geologic and other shocks in solids, display this behavior. In each of these fields the technical word "impedance" is used to describe how easily the wave passes through the medium in question. A boundary, then, is a place where the impedance changes, either quickly for hard collisions, or gradually for soft ones. The term "impedance mismatch" is often heard during discussions of reflection and transmission of waves or pulses. A big impedance mismatch between media, or at boundaries, causes big reflections.

These other fields of science and technology have contributed general knowledge useful in understanding musical sound. And vice versa, the early investigations of comparatively easily measured sound established the scientific basis for much of the subsequent studies of waves.

• QUESTION A.12.   Reread Section 8.2.1, "The Bell". Identify the word or group of words in that section that now could be replaced with the word "impedance." •

Appendix B.
Computer-Controlled Audio Electronics

The first use of audio electronics to improve acoustics was to make the sound loud enough to be heard by all the listeners. Microphones were placed near the sound source, audio amplifiers somewhere else, and loud speakers wherever needed. All frequencies were more or less changed by the same amount, although if the amplifier happened to have bass and treble controls, low and high frequencies could separately be made louder or softer. But, that was about the extent of any active acoustics.

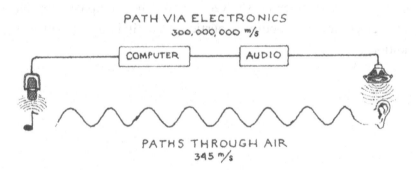

Of course, the sound arrived at the listener at two different times and often from different directions, but at least during pianissimo a listener at the rear of the hall heard something.

Although the theory of sound, its wave nature and how this occurs in air that consists of individually widely-spaced molecules was known by the end of the nineteenth century, the electronic audio technology that

Appendices

would make the detailed measurements of sound in rooms was not. As this developed the measurements provided a better idea of the acoustics of a room, and both the apparatus and the measurements became a part of what was available to change the acoustics: the same technology working hand-in-glove to reveal the details of acoustical problems and then being used to correct them.

This Appendix describes a part of this dual capability that has come from the integration of audio electronics and computers, and how this interplay can be used to change a room's acoustics. Certainly changing the room itself can alter a room's acoustics. Towards the end of this Appendix two QUESTIONS will ask you to suggest some improvements to a room's acoustical problems by using both room changes and Computer-Controlled Audio Electronics (C-CAE). This will be a genuine task for you, sometimes doable in several ways, sometimes doable but probably impractical, and sometimes not possible. In the meantime let's begin a more detailed look at C-CAE.

Figure 11.8, in Chapter 11, showed a way to use computer controlled audio electronics to modify acoustics. Here it is again with some additions.

Appendix B

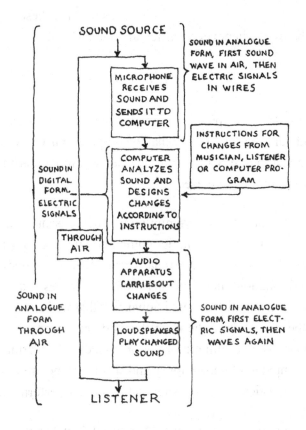

Figure B.1 Block Diagram of Normal and Computer-changed Sound From Source to Listener

We will concentrate on what is taking place in the boxes "COMPUTER ANALYZES SOUND AND DESIGNS CHANGES ..." and "INSTRUCTIONS FOR CHANGES ... ."Let's begin with the "INSTRUCTIONS ..." box.

The instructions could come from a person who might say something like "The pianissimo passages need more warmth ..." or might be the result of measurements that have detected differences between the actual and wanted acoustics. In either case these must be translated into computer statements that specify how each frequency must be changed. Or the computer itself might have been part of the apparatus

Appendices

that previously measured the room's acoustics; and these data are now ready to be used whenever a particular sound is wanted.

B.1 The Computer

Instead of going into ephemeral details of computer programming, we will spend our time specifying what the program must accomplish. And a look again at Figure B.1 reminds us that the computer is only a part of the apparatus that will be used.

Let's begin by making the assumption that the present acoustics of our room are known and stored in our computer's memory. By this I mean that the reverberation times and delay times of each frequency have been measured throughout the room, any room resonances and the locations of their nodes and antinodes are known, and that we have identified any areas in the room where the overall sound can become too soft to be heard. This is our baseline information. Once the performance begins both the computer's huge memory and speed will be used to create the particular acoustics wanted during that particular performance.

What is commonly called a computer's speed, how fast it can do things, depends on both the computer and whatever program it is executing. Computers do their basic operations in cadence with the ticking of an internal clock. In the dark ages of computing a reasonable clock rate was one million ticks a second (1 MHz or one megahertz). In the year 2009, 3000 MHz (3 GHz or three gigahertz) was possible; the same computer program will run 3000 times faster. In addition to faster clock rates, computer programs have become more efficient. The basic operations have been reprogrammed to take fewer ticks, and other operations that took many ticks streamlined or eliminated. This gained efficiency often makes the computer program much harder to read and understand, and therefore, harder to change when necessary. Programmers might want to use the added speed for their own convenience, and create more obvious, but longer, programs, which end up running no faster

Appendix B

then before. The computer program called a Fourier Transform is an example.

B.1.1 Analyzing the Sound

B.1.1.1 The Fourier Transform

Jean Baptiste Fourier (1786-1830) discovered that any waveform, such as the complicated ones you've seen on the oscilloscope, could be made from a superposition of single frequency waves. Earlier you combined two single frequency waves by superposition to form a complex wave. Fourier did the reverse. He started with a complex wave and wrote the recipes for finding what frequencies and how much of each of them was needed to form it. These amounts are called the Fourier coefficients. This process is called Fourier analysis, and the resulting listing of the necessary frequencies and their amplitudes is called the original wave's Fourier Transform. Although many, sometimes even an infinite number of waves might be necessary, it could always be done, and the sound of an orchestra can be gotten from a battalion of single frequency audio oscillators and loud speakers, each one changing the amplitude of its output in some complicated, but predetermined pattern. Synthesizers are already a step in this direction. Of course, Fourier had no idea of actually doing this, he just showed that it could be done and how. He wrote the recipes in mathematical forms that most clearly showed his basic ideas.

Almost from the beginning of modern computers people realized that using them to make Fourier Transforms was now a practical exercise even though it would require so much calculation and time. These problems, too, have been attacked and in 1965 J. W. Cooley and J. W. Tukey rewrote Fourier's equations in forms that required much less calculation[14]. In exchange for this, the ease of seeing the basic ideas was lost: an example of the trade off of understanding for increased efficiency, but the increased speed was irresistible. A computer program incorporating this improvement is called a Fast Fourier Transform (FFT).

Appendices

Whether or not we use Fourier's methods we can be guided by his ideas, which tell us that it is possible to modify a wave, no matter how complicated, by modifying the amounts of each of its component frequencies. We no longer have to consider the whole complicated wave, just its simpler independent Fourier component frequencies, one at a time.

B1.1.2 Sampling the Sound

The musical sound's waveform is continually and continuously changing as the music changes. The computer, on the other hand, operates in a discontinuous step-at-a-time following the ticking of its internal clock. How can the continuously changing music be represented and analyzed by the stop-and-start operation of the computer? You already know the answer. While you watch a movie you are seeing a series of still pictures presented to you, one after the other, about 1/20 of a second apart. Your eye, brain, and nervous system convert these individual pictures into continuous motion. If each frame lasted for a half second, you would be aware of the scene as a jerky sequence of individual pictures. So, in a similar way, if you hear discrete instances of a changing sound close enough together, you will "connect the dots" into a continuous flow of happening.

Computer clocks and programs are now fast enough to sample, analyze, modify, and replay musical sound so that you think it is continuous music. In fact, for digital recording of music, the problem is not in fooling your ear-brain-nervous system into perceiving the sound as continuous, but sampling it often enough so that none of the information about its high frequencies is lost.

For ordinary digital recordings the sampling rate is 40,000 samples/second. This may seem extremely fast, but is a leisurely task for a computer ticking three thousand million times a second (3000 MHz). Here is part of the graph of the electric signal from a complicated sound wave with the dots showing its sampled values.

Appendix B

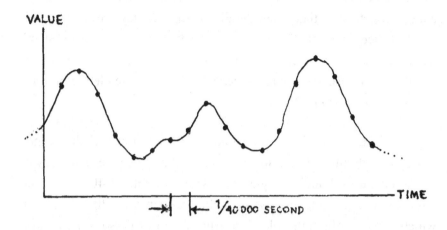

The musical sound wave is sampled every 1/40,000 second and the value of its loudness converted into a number, say between zero and four thousand. This process is called analogue-to-digital conversion (ADC). These numbers are stored in computer memory as a table of times and the digitized values at those times. Table B.1 shows part of these data.

TIME, s	DIGITIZED VALUE
•	•
•	•
1.000025	357
1.000050	386
1.000075	392
1.000100	365
1.000125	342
•	•
•	•

Table B.1 Sampling Times and ADC Values About One Second After Music Starts

Appendices

The digitized values are then written on a compact disc, spaced closely enough together so that when the disc is played they are read in their real time sequence. Both you and your hi-fi's audio "nervous system" have the inertia to convert these individual values back into continuous sound, a microscopic to macroscopic conversion called digital-to-analogue conversion (DAC).

Why does the sound have to be sampled 40,000 times each second? The rule of thumb is that it must be sampled at twice the frequency of the highest useful audio frequency. A sampling rate of 40,000/second indicates that the highest useful audio frequency is 20,000/second, which is above the threshold of hearing for most persons although it could be contributing to lower combination or complex frequencies or even to beats.

If the sampling rate is too low, say sampling a 2000 Hz wave 2000 times a second, the measured amplitudes would all be the same. You would be sampling at the same place in the wave every time and would get the same value. This is the same as sampling a straight horizontal line, and that is how the computer would register the waveform. The DAC would produce a constant value, which is not a sound wave. This is an extreme example of too low a sampling rate losing all the wave; but generally a too low a sampling rate will lose some of the frequencies present in an analogue wave.

While the sound wave is in the digital form it can be altered by simply adding to or subtracting from the stored numerical values, easy and quick work for a computer. However, doing this means that you are changing the shape of the whole wave all at once, and it may not be simple to determine which of the values in the above table should be changed, or by how much, if you just wanted to change the amount of particular frequencies. Having the computer do an FFT will take more computing time, but the constituent frequencies and their amplitudes would be ready for a conceptually more direct adjustment. Our table grows additional columns, one for each of the constituent frequencies.

Appendix B

The workload of our computer and the amount of memory needed to store the results have increased.

What makes the FFT process more complicated than ADC is that the FFT must have an interval of the music to work with, and the length of this interval must be at least as long as the period of the lowest constituent frequency. For example, if you want constituent frequencies as low as 50 Hz, the FFT must have about 1/50 second of music to analyze. This reciprocal relationship between wanted frequency and necessary sample length is universal and fundamental; this means that even though each sample can begin much more often, you cannot expect to get the information into each line of the columns before at least 1/50 second, or 0.020 second, after the sampling begins. The electronically generated delay times will be at least this long, but this is within the absolute limit of 0.060s after which echoes occur (see Chapter 11, Section 11.1.2.3).

So, the FFT process is not the normal ADC described above. Whereas the ADC simply calculates a number telling the height of the total music's oscilloscope trace 40,000 times a second, the FFT takes an interval of music 1/50 second long and calculates the Fourier Transform for the frequencies and their amounts for that 1/50 second interval. It does this again and again, one after another as the music plays. The musical piece is divided into many short, 1/50 second, intervals and an FFT is done for each one of these. If you want this frequency-amplitude information generated more often then every 1/50 second, you must have a second computer that begins its 1/50 second interval while the original computer is taking its interval. For example, if the second computer started its sample at the mid interval of the original's, you would get frequency-amplitude information updated every 1/100 second. This scheme of using several computers together is a type of **parallel computing**. Another type is explained below.

Recall that delay times should be between 0.035 s and 0.060 s. For an FFT analyzed and then adjusted sound wave this will be the time allowed from the start of a sample to when the correction is played and heard. Let's try to find out if this is enough time for the computer and the audio electronics to do their jobs. Assume that we do want the FFT to include a 50 Hz frequency and that the listener is 5m away from a corrections-generating loudspeaker and 10m away from the direct sound source. So, the computer has the time it takes the direct sound to reach the listener plus the delay time, and minus the sampling interval time and the time it takes the corrections from the loudspeaker to reach the listener, to do its work. For a 0.035s delay time these are 0.028 s + 0.035 s - 0.020 s - 0.014 s which equals 0.029 s. Not very long, but it is 87 million ticks of the computer's 3000 MHz clock. Is this enough? Maybe.

• QUESTION B.1. Here is the longitudinal cross section of a small auditorium. It has a computer-controlled audio system used to shape the sound according to the players' wishes. The microphone and the loudspeakers are shown.

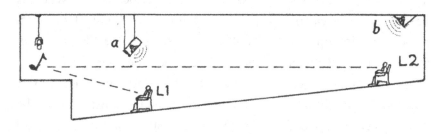

DISTANCES:

SOUND SOURCE-MICROPHONE :   2.0m
SOUND SOURCE- LISTENER L1 :   5.0m
SOUND SOURCE- LISTENER L2:   15 m
LOUDSPEAKER a LISTENER L1:   7.0m
LOUDSPEAKER b LISTENER L2:   7.0m

You want the sound from the loudspeakers to reach the listeners within 0.035 s to 0.060 s after they hear the direct sound. The computer uses an FFT program with a sample length long enough to include a constituent frequency of 50 Hz.

Appendix B

How much time does the computer have to design the sound corrections for each listener?

Could one set of computer generated delay times satisfy both listeners L1 and L2? Why? •

The quest for increased computer speed has not been limited to making computing more efficient. Other ways include:

• PARALLEL COMPUTING. In addition to the parallel computing scheme described above, there could be one in which the computations, such as finding the frequencies and their amplitudes, are not done using an FFT computer program, but having many computers working simultaneously, each one finding the amplitude of one pre-specified frequency.

• HARD-WIRED OPERATIONS. Repeatedly done operations are performed in separate parts of the computer that only do that one operation and do not need the flexibility to accommodate a changeable computer program. They have their instructions soldered into them and are released from needing the synchronization a clock provides. They spend no time waiting for a computer program to make decisions about what to do next. Hard-wired operations take one millionth of a second (0.000001 s) or less.

It is possible, and may be necessary, to combine and mix all these techniques to increase computer speed.

During a musical performance, microphones throughout the room will sample the sound and, based on a comparison of it with the baseline information and what acoustics are wanted, the computer will specify the sound that must be played by the loud speakers. The computer applies a set of corrections, but only when and as the sampling reveals that they are needed: active acoustics.

Appendices

The collection of the room's baseline acoustics can be a leisurely activity; the real time sampling and correcting the sound during a performance has the severe time constraints mentioned earlier. Here is a block diagram of this process, also showing how it fits in the complete process more generally than shown in Figure B.1.

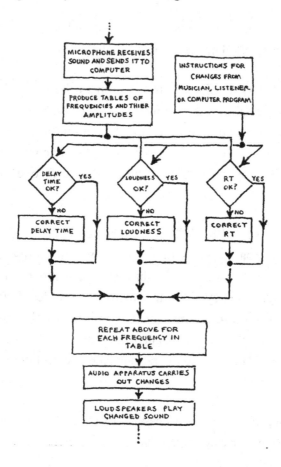

Figure B.2 Block Diagram of Computer Making Corrections to the Sound

The box "PRODUCE TABLES OF FREQUENCIES ..." describes calculations that might be done by a computer FFT, or operations that might be done using a computer and analogue electronic circuits called **bandpass filters**. Here's how this latter method works. A collection

Appendix B

of bandpass filters also separates musical sound into its constituent frequencies. Figure B.3 is a sketch of a bank of bandpass filters, each followed by its ADC, and then connected to the computer, which will calculate the necessary corrections for each frequency. The audio input and output are also shown. Note also where the music changes back and forth from analogue to digital format.

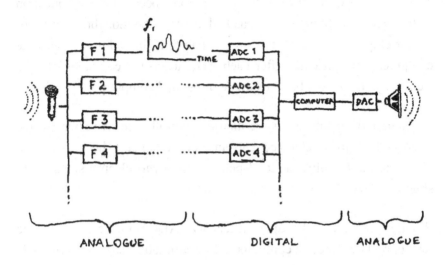

Figure B.3 Bandpass Filters Separate the Frequencies

Each bandpass filter, F1, F2, ... , transmits only one, or more realistically, only a very narrow band of frequencies. The above figure shows a bank of these filters doing a similar, but not identical, job as an FFT. For one thing, the FFT selects the constituent frequencies itself, whereas you can choose the number of bandpass filters and each one's frequency. The FFT selects a satisfactory set; you might leave out an important one. The FFT calculates the frequencies and their amplitudes, i.e. constructs the Fourier Transform, and presents this information in digital format, a series of numbers already in computer language. The frequency of each bandpass filter is chosen before hand and hard-wired into its circuitry. The output of each bandpass filter is an analogue signal not yet in digital format. An ADC must be performed for each output, and then the computer can

Appendices

use the digitized values. If you combined the outputs of each bandpass filter again by connecting together their output wires, at a point before the ADCs, the original analogue signal would appear again. The system shown above has increased speed because all the bandpass filters are in parallel working simultaneously. There must be a bandpass filter and an ADC for each frequency you want and this requires additional circuitry. That's another of the trade-offs for increased speed. Such a duplication of the hardware, though, is not as bad as might be thought. Integrated-circuit chips for doing all this are available. If, in addition, a separate computer were provided after each ADC to do the corrections for its frequency, that would be an example of parallel computing.

No matter what kind of wave analysis you choose, the sound wave must first of all be changed into an electric signal. The microphone does this. And finally, the loudspeakers reconvert electric signals into sound waves.

Whether an FFT or bandpass filters and ADCs do the analysis, the results will be stored in computer memory in tables such as Table B.2.

	$f1$	$f2$	$f3$	• •
$t1$	$AMPLITUDE_{11}$	$AMPLITUDE_{21}$	$AMPLITUDE_{31}$	• •
$t2$	$AMPLITUDE_{12}$	$AMPLITUDE_{22}$	$AMPLITUDE_{32}$	• •
$t3$	$AMPLITUDE_{13}$	$AMPLITUDE_{23}$	$AMPLITUDE_{33}$	• •
$t4$	$AMPLITUDE_{14}$	$AMPLITUDE_{24}$	$AMPLITUDE_{34}$	• •
$t5$	$AMPLITUDE_{15}$	$AMPLITUDE_{25}$	$AMPLITUDE_{35}$	• •
•	•	•	•	• •
•	•	•	•	• •

Table B.2 Sampling Times and Frequency Amplitudes

The FFT or bandpass filter-ADC methods give the same type of tables of times, frequencies, and loudnesses, all ready to be modified. Both

Appendix B

require large amounts of computer memory to store all these numbers, and fast computational speed to make use of all of them.

Compare Table B.2 to Table B.1, which shows the output from a simple ADC of the total wave. Table B.1's single column of digitized values has grown considerably, adding a column for each frequency. The FFT or bandpass filters and ADCs methods give us the frequencies and their amplitudes all ready to be modified. In exchange for this they require more computer work, more memory for storage of the results, and possibly more electronic hardware for the bandpass filters and ADCs.

This takes us through the box "PRODUCE TABLES ..." in the block diagram in Figure B.2. Now we will look at how the decisions and resulting corrections are made that will change the loudness and reverberation times and time delays.

B.1.2 Modifying the Sound

B.1.2.1 Changing Loudness and Reverberation Times, Reflected Sound

Insufficient loudness can be due to a local dead spot for a particular frequency or because all the sound there is too soft. The loudspeakers might be the only source of sound of that frequency available there, and, if so, should be placed pointing to the listener from the location of the sound source so that they give the impression of producing direct sound. The sound from these speakers, wherever they are, will be part, or all, of the direct sound and must arrive at the listener before any reverberant sound. Otherwise the listener would be looking at a sound source but hearing sound coming from somewhere else. I realize that this is the situation while listening to music through earphones. In fact, people sometimes describe good stereo music as coming from inside their heads, but this is not the kind of perception I am trying to achieve: a mostly direct and natural sound.

Appendices

Let's look at a case where one frequency is not loud enough.

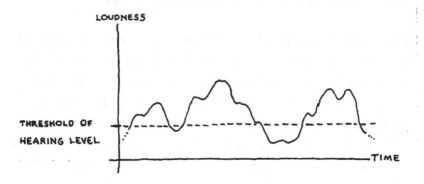

The above graph is a smoothed curve drawn from the values in one of the columns in Table B.2. It shows that the loudness of that frequency drops below the threshold of hearing or the general noise level and is too low to be heard some of the time. The listener will hear this frequency come and go instead of being louder and softer. The computer will scan these numbers and compare them to a typical threshold of hearing or to the general noise level, and order the audio system to add sound as needed through loudspeakers. However, doing this only to those amplitudes below the threshold of hearing or noise level would decrease the music's dynamic range (the difference in loudness between the loudest and softest tones), and, because it might be dealing with only a few of the many frequencies present, might also change the timbre. Once the deficient amplitude has been identified the computer should add the loudness correction to the whole sound, i.e., turn up the volume of the loud speakers. How much? Probably not enough to achieve the loudness the musicians hear, but certainly enough, or just enough, to get rid of the room's bad seats.

Loudness and reverberation time are related: louder sounds take longer to die away. Figure B.4 shows the decay of two tones, A and B, having identical frequencies but different initial loudnesses.

Appendix B

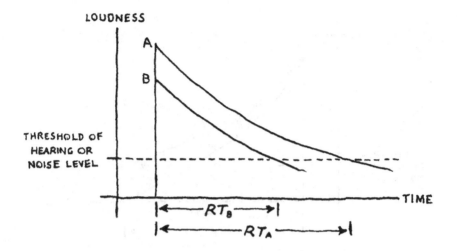

Figure B.4 Initial Loudness and Reverberation Times

Tones A and B begin at the same time, but because they have different initial loudnesses, they have different reverberation times. Exactly how long each is depends on the amount of reflection when the sound hits an obstacle, and, therefore, on the construction of the room. Appendix A discusses what happens when sound encounters obstacles; but generally weak reflections cause shorter reverberation times, and vice versa.

Loudspeakers can be used to mimic reflections, and reverberation times can also be lengthened without changing the loudness of the initial sound, by adding sound during the decay. The graph below shows how sound is added.

Appendices

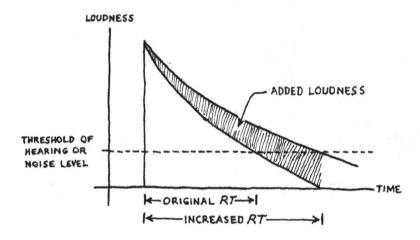

Figure B.4.A Added Loudness to Cause Longer Reverberation Time

The shaded area in Figure B.4.A indicates how much loudness must be added. Note two things: first, the amount of loudness in Figure B.4.A that must be added depends on the loudness of the attack, the initial sound; and second, the added loudness is added to both the audible and inaudible sound. The inaudible sound waves are still there; it is the limitations of your hearing that makes them inaudible. They were just too quiet for you to hear them before. The computer identifies the need for a longer reverberation time and calculates the amount of correction needed. It then increases the values of the entries in the columns of Table B.2 after the initial loudness, and the loudspeakers plays the extended sound. Let's see how this can be done.

The required reverberation time for each frequency will have been put into the computer program and are part of the instructions in box "INSTRUCTIONS FOR CHANGES FROM …" in Figures B.1 and B.2. This data most likely will have been entered before the musical performance, but could be changed while the performance is under way. During the performance the computer scans the latest row of frequencies

Appendix B

and their amplitudes in Table B.2, and when it finds an attack determines the initial loudness and calculates the needed reverberation time correction for each frequency. The computer generates the instructions to do this, which are then sent to the electronic audio system to play. Figure B.5 shows a block diagram of this process.

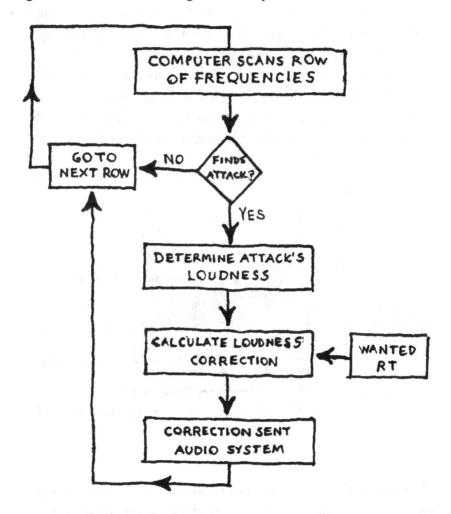

Figure B.5 Block Diagram of Computer
Correcting Reverberation Times

Because the *loudness of the initial sound* is the loudness of the *direct sound* (it arrives first), and *reverberation time* is a measure of how

- 253 -

Appendices

the loudness of the *reflected sound* changes, it is possible to adjust them independently. An increase of the initial loudness will certainly increase the reverberation time (see Figure B.4), but it is also possible to increase the reverberation time without increasing the initial loudness (see Figure B.4.A). Loudness can be changed by turning the volume of any direct sound loudspeakers up or down, or by adding or removing reflecting surfaces behind the musicians, or by moving the musicians nearer or farther away from the listeners, or even by playing the music louder or softer. Once a satisfactory loudness is achieved, the reverberation time can then be set. Here is a graph showing the results of incorporating some of these possibilities into Figure B.4.A. Both increased and decreased reverberation times are shown.

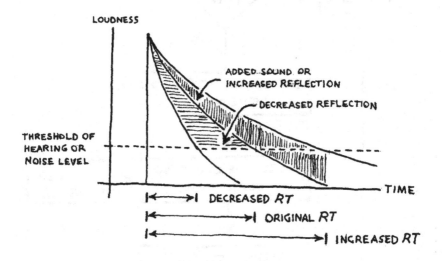

The reverberation time is increased by adding loudness from computer-controlled loudspeakers, or by increasing the amount of sound reflected from the room's boundaries. It is decreased by lessening the reflection. Changing reflection requires changing the physical structure of the room, and would not be done while the music plays. If the room could be reconfigured in a few minutes, say in the time between musical

Appendix B

selections, a computer could speed up this job by providing menus of possible modifications and their effects, or could even measure the existing reflections and design a room configuration to improve them. A more practical idea might be to begin with a room having a minimum reverberation time and use computer-controlled audio to produce longer reverberation times by adding loudness during the decays.

• QUESTION B.2.   You are the audio engineer for an auditorium that has a complete set of computer-controlled audio electronics throughout the hall, and can have its physical properties changed to vary reflections.

It takes a day to change the room's physical properties.

Discuss what you would recommend in the following situations if you were told two days before the concert that:

> The music's direct sound loudness is satisfactory, but the reverberation time (RT) is too short.
>
> The music's direct sound loudness is satisfactory, but the RT is too long.
>
> The music's direct sound is too loud, and the RT is too short.
>
> The music's direct sound is too loud, and the RT is too long.
>
> The music's direct sound is not loud enough, and the RT is too short.
>
> The music's direct sound is not loud enough, and the RT is too long. •

• QUESTION B.3.   Same as the above, except that now you are told the auditorium's defects an hour before the concert.•

Appendices

• QUESTION B.4.   A preferred way to begin designing a process, whether it will be a computer program or the manufacture of a bicycle, is to use block diagrams. Figures B.1, B.2, and B.5 are examples. The process generally flows downward from the top block with the lines and arrowheads indicating the way. Rectangular blocks indicate jobs to be done and diamond-shaped ones show decisions. As the design evolves one block may become a set of them showing the details of the process.

Part of the skill needed to block out a process is knowing what operations and apparatus are already available and can be just named in a block, and when it is necessary to provide block to block details of what is to be done next. For example, bandpass filters and ADCs are "off the shelf" items and therefore Figure B.3 doesn't need more blocks.

Figures B.4 and B.4.A show how added loudness causes added reverberation time. It has been suggested that a useful loudness correction would be to add enough loudness to get a satisfactory reverberation time.

Show how would you fit the block diagram Figure B.5 into the block diagram Figure B.2 in order to do this. This may be an easy "cut and paste"; or it may not. •

B.1.2.2 Delay Times

Delay times were discussed earlier. If they are too short, the computer-generated sound can be delayed: stored in its memory for a while, before being played. If they are too long, and mostly due to the sound from loudspeakers, and if more efficient computing cannot make up the difference, they must be attacked at their cause. The long reflection paths must be shortened with intermediate reflectors, or destroyed with more absorbent obstacles. Appendix A describes these possibilities.

Appendix B

B.2 Disclaimer

Writing about the capabilities of computers and computing is writing history; and this Appendix should have a "pull date" although it's already past. I have tried to prolong its life, and also not talk about things I don't know much about, by staying away from particulars. One exception was the use of year 2009 computer speeds to show that it is already possible to do live real-time sound modification, which I call active acoustics.

Appendix C.
Science Symbols: Physical Quantities, Physical Objects, And How They Are Written

Science and music are sometimes thought to be almost exclusive, but they are not. Both use written symbols to represent some quite complicated things. A musical example is the staff with its notes, clefs, and sharps and flats. A musician learns these symbols well enough to "be able to hear" the music by reading the score. Musical notation, once it is understood, is an economic and elegant system of symbols for presenting musical sound.

Science, too, has its written symbols. They represent Physical Quantities and Physical Objects[15].

Physical Quantities have dimensions such as length, time interval, temperature, pressure, or mass. The symbols for Physical Quantities are usually one or two letters, often the initial ones in their names. These symbols are written in italics. The value of a Physical Quantity is written as a number with the Quantity's units. The values are not written in italics. Examples of Physical Quantities and their values are shown in the following Table. Metric system units are used.

Appendices

PHYSICAL QUANTITY	SYMBOL	UNITS	EXAMPLE OF VALUE
frequency	f	1/s, Hz	440/s, 440 Hz
period	T	s	0.0022 s
wavelength	WL or λ	m	1.5 m
pressure	p	kg/m s², Pa	1×10^5 kg/m s²
temperature	T	K or °C	293 K or 20 °C
loudness		dB	93 dB
equivalent loudness		phon	110 phon
volume	V	m³	1.35 m³
mass	m	kg	0.52 kg
length	l or L	m	9.0 m
speed	v	m/s	345 m/s
reverberation time	RT	s	0.23 s

Table of Some Physical Quantities with their Symbols, Units and Examples of their Values

The symbol for a Physical Object is also an abbreviation of its name. Physical Objects can have size, but not a numerical value with metric units. The symbols for Physical Objects are not italicized.

The following Table shows examples of Physical Objects and their symbols.

PHYSICAL OBJECT	SYMBOL
piano key or its tone	A, C, G, …
wavefront	WF
musical interval	PERFECT FIFTH (P5), MINOR THIRD (m3)

Appendix C

disc	D
analogue-to-digital converter	ADC
analogue-to-digital conversion	ADC
band pass filter	F
Fast Fourier Transform	FFT

Table of Some Physical Objects with their Symbols

The symbols for Physical Quantities and Physical Objects can have subscripts and superscripts attached, and numbers, often integers, added. These give more information about the Quantity or Object. For example, symbols for frequency, period, or wavelength have number subscripts added to make them the frequency, period, and wavelength of a particular modes: f_1 is the frequency of the 1st mode; T_2 is the period of the 2nd mode; WL_3 or λ_3 is the wavelength of the 3rd mode. Subscripts can also specify the type of cavity: f_{cyl} is a frequency of a cylinder; f_{cone} is the frequency of a cone. The symbol f_{do} is the frequency of some pitch do. The symbol A_4 represents the 49th piano key (see Figure 2.2 in Chapter 2).

An integer is placed behind the symbol for a Physical Object to specify a particular object in a group of them: WF1 and WF2 are two wave fronts. Band pass filters F1 and F2 are different because each passes a different band of frequencies. Both of these filters might be the same kind of instrument, but with their dials set for different frequency bands.

A number placed in front of the symbol for a Physical Quantity multiplies the value of that Quantity: $2WL_1$ is a wavelength twice as long as WL_1. The equation

$$WL_2 = 3WL_1$$

states that the wavelength of the 2nd mode of oscillation is three times longer than the wavelength of the 1st mode.

Answers To QUESTIONS

CHAPTER 1

QUESTION 1.1: If sound were high-speed invisible particles, would you be able to hear around corners or behind screens? No, not unless there was something present to bounce them around or behind. This suggests doing an experiment in an open area where nothing is present to cause bounces. The particle model would look bad if you could still hear around corners or behind screens.

CHAPTER 2

QUESTION 2.1: The graph in the hint does show a zero pressure level. It's the horizontal axis. Both graphs present the same information, but because the pressure fluctuations shown on the first graph are so big, its zero pressure level must be way below the horizontal axis.

QUESTION 2.2: A crest acts like a convergent lens. Compare the light rays through the glass lenses to the water crests and troughs.

QUESTION 2.3: At locations where equal amounts of high and low pressures meet, the excess concentration of air molecules in the high will fill in the deficiency of air molecules in the low, and atmospheric pressure will result. If this situation continues your ears will always be in a steady atmospheric pressure and you won't hear the sound.

Answers to QUESTIONS

QUESTION 2.4: Draw a horizontal line at the level of atmospheric pressure. When the graph is above this line the sound wave's pressure is higher than atmospheric pressure. Below this line the pressure is lower than atmospheric pressure, i.e., partial vacuum. When the graph crosses the line, the sound's pressure is atmospheric pressure, but only at those instants. Pick a few of each of these situations and read their times from the time axis.

For example:

higher than atmospheric pressure: 0.0018 s, 0.0023 s

atmospheric pressure: 0.0015 s, 0.0031 s

lower than atmospheric pressure: 0.002 s, 0.0033 s

No, this sound is not noise. It has repeatable cycles.

period: about 0.00067 s

frequency: about 1500 cycles/s

wavelength: about 0.23 m

QUESTION 2.5: Refer to Chart 2.1 and Figure 2.3.

KEY	do	re	mi	fa	sol	la	ti	do'
G Major	G	A	B	C	D	E	F#	G
G# Major	G#	A#	C	C#	D#	F	G	G#
G# Minor	G#	A#	B	C#	D#	E	F#	G#

QUESTION 2.6: Refer to Chart 2.1 and Figure 2.3.

KEY	do	re	mi	fa	sol	la	ti	do'
GIVEN	D	E	F	G	A	B	C	D
D Major	D	E	F#	G	A	B	C#	D

Answers to QUESTIONS

QUESTION 2.7: Refer to Chart 2.1 and Figure 2.3.

KEY	do	re	mi	fa	sol	la	ti	do'
GIVEN	D	E	F	G	A	B	C	D
D Minor	D	E	F	G	A	A♯	C	D

QUESTION 2.8:

The wavelength decreases by a factor of 4.

QUESTION 2.9: MAJOR SIXTH   575 Hz

QUESTION 2.10:   mi : 438 Hz

fa : 467 Hz

sol : 525 Hz

QUESTION 2.11: For example, pick piano keys numbered 4, 40, and 76 for your dos, i.e., C_1, C_4, and C_7.

For a PERFECT FIFTH, $f_{do} / f_{sol} = 2/3 = 0.667$, from Table 2.1, and according to Figure 2.2, $f_{do} / f_{sol} = f_{C1} / f_{G1} = 32.703$ Hz $/ 48.999$ Hz $= 0.667$, good agreement.

For $f_{do} / f_{sol} = f_{C4} / f_{G4} = 261.63$ Hz $/ 392,00$ Hz $= 0.667$, also good agreement.

Try C_7: $f_{do}/f_{sol} = f_{C7}/f_{G7} = 2093$ Hz $/ 3136$ Hz $= 0.667$; continuing good agreement.

Answers to QUESTIONS

For MAJOR THIRDS, $f_{do}/f_{mi} = 0.800$, from Table 2.1,

and $f_{C1}/f_{fE1} = 0.794$; and $f_{C7}/f_{E7} = 0.794$; not bad agreement.

And etc.

CHAPTER 3

QUESTION 3.1:   Use equation (2.3).

speed of wave = 18 m/s

average speed of dash winner = 10 m/s, about.

QUESTION 3.2:

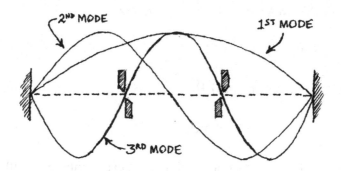

Two good locations shown.

QUESTION 3.3: Here is the sketch of the construction of the complex wave for the superposition of the 1st and 3rd modes completely out of phase.

Answers to QUESTIONS

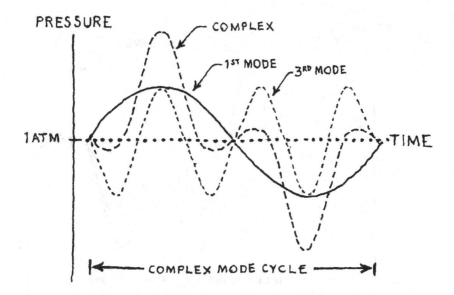

And the oscilloscope display of this complex wave is

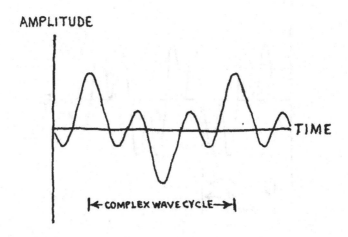

Compare the above two graphs to those for the in phase superposition of the 1st and 3rd modes:

Answers to QUESTIONS

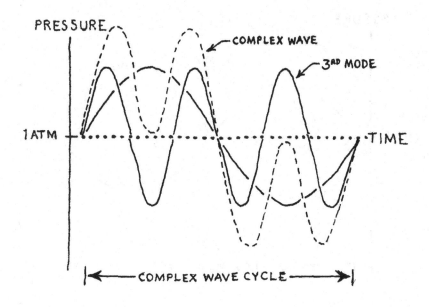

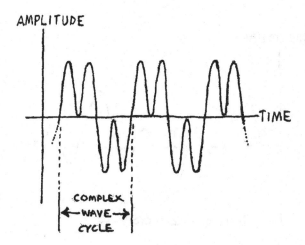

Wavelength of complex wave is the same as for the in-phase complex wave. You would hear the same pitch.

Answers to QUESTIONS

QUESTION 3.4:

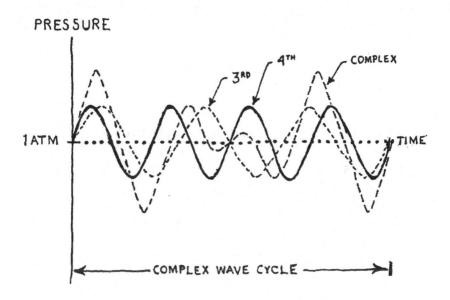

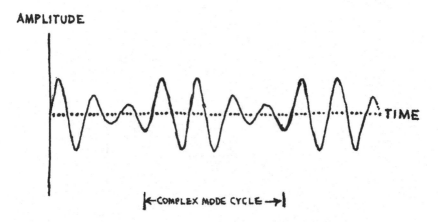

The period of the complex wave is 3 times the period of the 3rd mode and 4 times the period of the 4th mode.

Answers to QUESTIONS

TYPE	MODE	INTERVAL	
Complex	1	⎤ OCTAVE + PERFECT FIFTH ⎤	
Played	3	⎦	2 OCTAVES
played	4	⎤ PERFECT FOURTH ⎦	

You will examine this interval again in Example 2.) of Section 3.7.

QUESTION 3.5:

TYPE	FREQUENCY Hz	INTERVAL	
Complex, $f1$-$f2$	50		
Played, $f2$	150	⎤ PERFECT FOURTH	
Played, $f1$	200	⎦	OCTAVE ⎤
$2f2$	300	⎤ PERFECT FIFTH ⎦	
$f1+f2$	350	⎤ PERFECT FOURTH	OCTAVE
$2f1$	400	⎦	⎦

Same intervals. The frequency of the fundamental is 50 Hz, an octave lower pitch than Examples 1.) and 2.).

QUESTION 3.6:

N periods of 500 Hz = (N+1) periods of 612 Hz. Recall that period = 1/ frequency, and so the period for 500 Hz = 1/500 Hz = 0.00200 s. Similarly, the period for

Answers to QUESTIONS

612 Hz = 0.00163 s. Thus,

N (0.00200 s) = (N+1) (0.00163 s) = N (0.00163 s) + 0.00163 s, or

N (0.00200 s - 0.00163 s) = 0.00163 s, and

N = 4.41.

QUESTION 3.7: The answer to this QUESTION requires you to use most of the techniques already presented in Chapter 3. In that sense it is a review question about information you will need later.

If the frequencies were 100/s and 201/s the answers to the above parts would not change much, probably not enough to hear the difference. But, beats might now be present between the 2f combination frequency and the 201/s tones. If so, beats can be created in the ear-brain system. I don't believe this is yet established.

QUESTION 3.8: The answer requires a lot of calculator work. For a pressure of 0.1 Pa,

dB = 20 $\log_{10}$(0.1 Pa/2x10^{-5} Pa) = 74 dB, and so forth.

Phons and dBs have the same values at 1000 Hz. At other frequencies they deviate in value from each other. Phon measurements are subjective and mimic the non-linear complexities of the ear-brain system. dB measurements are made with scientific instruments.

CHAPTER 4

QUESTION 4.1: Find that f_{do}/f_{mi} = 4/5, directly from Table 2.1. So,

$$f_{do}:f_{mi}:f_{sol}:f_{do'} = 4:5:?:?.$$ But,

$$f_{mi}/f_{sol} = (f_{do}/f_{sol})(f_{mi}/f_{do}) = (f_{do}/f_{sol})(1/(f_{do}/f_{mi})) =$$

Answers to QUESTIONS

(2/3) / (1/(4/5)) = 5/6, and now you know that

$$f_{do} : f_{mi} : f_{sol} : f_{do'} = 4 : 5 : 6 : ?\,.$$

Now just recall that $f_{do}/f_{do'} = 1/2 = 4/8$, an octave; and finally

$$f_{do} : f_{mi} : f_{sol} : f_{do'} = 4 : 5 : 6 : 8$$

CHAPTER 5

QUESTION 5.1:   Equation (3.2),

$$f_n = v\,n\,/2L,$$

with v and n unchanged (same string, same mode) shows that if L gets bigger, f_n gets smaller. Doubling L will halve f_n, and half the original frequency is a pitch an octave lower. Note that this argument will work for any constant value of n. All the modes' frequencies become an octave lower.

QUESTION 5.2: How many octaves in a piano's range? Look at Figure 2.1 and count them. There are over 7. Call this 8, it's close enough and we don't want to shortchange the number of keys. According to QUESTION 5.1 doubling the string's length will cause the string's pitch to lower by one octave. We need 8 doublings. Since

$$2^8 = 256$$

the string must change length by a factor of 256. Of course, it doesn't and other methods are used to change its vibrational frequency.

Answers to QUESTIONS

QUESTION 5.3:

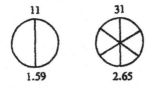

So,

$$f_{11}/f_{31} = 1.59/2.65 = 0.600.$$

Consult Table 2.1.

CHAPTER 6

QUESTION 6.1:

TEXT	YOUR
123	321
132	312
213	231
VS.	
231	213
312	132
321	123

Answers to QUESTIONS

CHAPTER 7

QUESTION 7.1: Pick a point not corresponding to points a or b. Use the same method shown and find that at your point the maximum displacement of the standing wave will be less than at point a.

QUESTION 7.2:

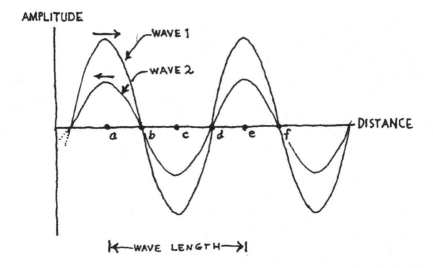

WAVE 1 is moving to the right, WAVE 2, to the left. They are crossing as shown at some instant. Both WAVEs have the same wavelength (and frequency), but their amplitudes are different. Because of this there are no distances where the resultant wave's displacement is always zero. There are distances where the amplitude of the complex wave will always be largest and other distances where the amplitude will always be smaller, but no real nodes. The complex wave is changing displacement everywhere, but its amplitude is never zero

Answers to QUESTIONS

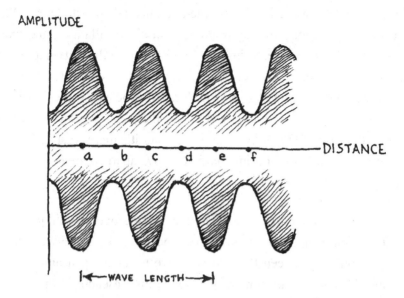

Only the envelope of the resultant complex wave is shown above. The shaded area within the envelope represents the possible values of the displacement there at various times. Such shading for standing waves created by intersecting equal amplitude traveling waves would look like this:

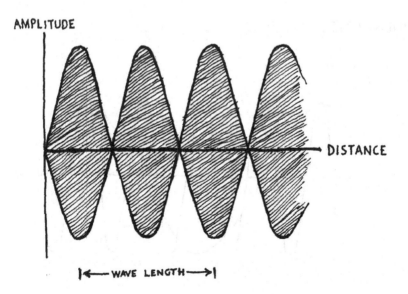

Answers to QUESTIONS

For the complex wave formed by the two unequal amplitude traveling waves, the locations of maximum amplitude oscillations are fixed at distances a, c, and e. Minimum amplitude oscillations happen at distances b, d, and f. Distances a, c, and e can be called locations of antinodes, but there are no real nodes. Nonetheless, these locations are fixed; there is a standing wave. Not the same as you're used to seeing, because of the lack of real nodes, but it's there. Just as for any standing wave the neighboring maxima are $\lambda/2$ apart, as are neighboring minima.

Of course, in order to obtain standing waves, two traveling waves must be going in opposite directions and have the same frequency. But, as you've just seen they do not have to have equal amplitudes. In musical instruments some of the wave must "escape," either out of the cavity or into the soundboard or soundbox, and, finally, into the air to be the sounds you hear. Some mechanism must be present to keep the amplitudes of the WAVEs about equal, and thus produce a strong resonance with definite nodes and antinodes. Otherwise, as you've just read, the pressure displacements in the envelope will diminish and so will the resonance. The instrument will quit playing.

QUESTION 7.3:

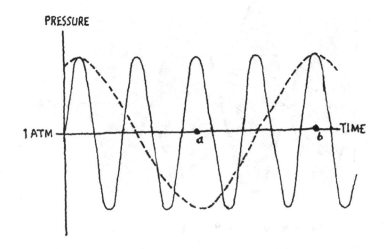

Answers to QUESTIONS

The higher frequency mode is not getting the puff it needs at time a, because the lower frequency mode is low there then and the two modes add to atmospheric pressure. The reed will not open, and the higher frequency mode goes without a puff for a cycle. It dies away. The lower frequency mode doesn't care about this. It doesn't need a puff at a; in fact it doesn't need one until time b.

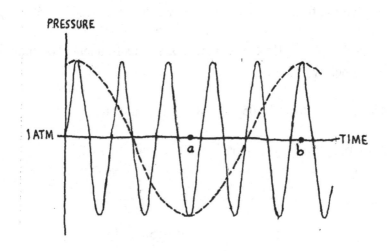

Compare what's happening at time a here and in the previous graph.

QUESTION 7.4:

cylinder and reed:
$\begin{cases} f_1 = 58/s \\ f_2 = 174/s \\ f_3 = 290/s \end{cases}$
clarinet, some organ pipes

cylinder blown:
$\begin{cases} f_1 = 115/s \\ f_2 = 230/s \\ f_3 = 345/s \end{cases}$
flute, pan pipes, some organ pipes

cone and reed:
$\begin{cases} f_1 = 115/s \\ f_2 = 230/s \\ f_3 = 345/s \end{cases}$
bassoon, oboe, saxophone, recorder, some organ pipes

Answers to QUESTIONS

CHAPTER 8

QUESTION 8.1:

From Figure 2.2 find C_3 = 130.81 Hz. This is f_1: and f_7 = 7(130.81 Hz) = 915 6 Hz.

Recall that the frequency multiplier between semitones is 1.05946.

So, x (1.05946) = 915.6 Hz, where x is the frequency of the first semitone below 915.6 Hz.

And so, x = 915.6 Hz / 1.05946 = 864.2 Hz; and etc.

Look up piano frequencies in Figure 2.2., and find that $7f$ lies between A_5 and $A_5\#$.

A_5=880/s
7f=915.6/s
$A_5\#$=932/s

Now prepare a table showing the six semitones below $7f$ and the nearest piano key:

semitones below 7f	piano
864.2/s	880.0/s, A_5
815.7/s	830.6/s, $G_5\#$ or $A_5\flat$
769.9/s	784.0/s, G_5
726.7/s	740.0/s, $F_5\#$ or $G_5\flat$

Answers to QUESTIONS

685.9/s	689.5/s, F_5
647.4/s	659.3/s, E_5

QUESTION 8.2:

n = 1 , first mode.

So, f_{cyl} = (345 m/s) / 4L = 1/2 (345 m/s)/ 2L = 1/2 f_{cone} , or

$2 f_{cyl} = f_{cone}$.

QUESTION 8.3:

For a cylinder, the n for the 3rd mode is 5.

For a cone, the n for the 3rd mode is 3. So,

f_{3cyl} = (345 m/s) 5/4 L, and f_{3cone} = (345 m/s) 3/2 L . Or,

4/5 $f3_{cyl}$ = 2/3 f_{3cone} , and finally

$12/10 f_{3cyl} = f_{3cone}$.

QUESTION 8.4:

MODE	f_{cone} / f_{cyl}
1	2
2	4/3=1.3
3	12/10=1.2

Answers to QUESTIONS

These ratios are approaching 1. The frequencies of each higher mode of the cone and cylinder are becoming equal.

QUESTION 8.5: The 16th mode is an octave above the 8th mode.

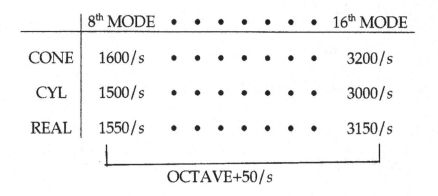

$3150/s / 1550/s = 2.03$.

The ratios of the frequencies are: 2.14, 2.07, 2.03, ...

A ratio of 2.00 is not sharp at all.

QUESTION 8.6:

Radius of area A = 5 mm

Radius of area at WL1 = 7 mm

Radius of area at WL2 = 13 mm

Area A = π (25 mm^2)

Area at WL1 = π (49 mm^2)

Area at WL2 = π (169 mm^2)

Ratio of areas from A to WL1 = about 2

Answers to QUESTIONS

Ratio of areas from A to WL2 = 6.8

The cross sectional area for WL1 increased by about 100%, almost doubled.

The cross sectional area for WL2 increased 580%, almost 7 times bigger.

So, the smaller cross sectional area change was for WL1, and it would have to continue farther down the bell before being reflected. WL2 has a bigger area change one wavelength down stream from area A and it will reflect at area A. Remember that longer wavelengths have lower (smaller) frequencies, and so the bell makes the length of the horn longer for the higher frequency of WL1. This lessens this frequency and makes it more flat. Just what you need to counter the 50Hz that our horn's octaves were too sharp.

CHAPTER 9

QUESTION 9.1: The explanations of standing waves in closed-end cylinders made in Section 7.2 of Chapter 7 will get you started answering this QUESTION, but will fall short of describing real clarinets. This should be no surprise; clarinets were not already "there" waiting to be discovered like DNA molecules. They were created and developed by craftsmen over several hundred years. Learn, and even admire, the general rules presented in *TSMS*, but realize that in order to understand real clarinets you must study real ones.

Let's begin with a simple closed-end cylinder and add the reality of real clarinets as necessary. If you reach a point where the original simple ideas are no longer justified, or useful as starting points, you must decide what to do next.

Answers to QUESTIONS

Begin answering this QUESTION by applying the explanation of standing wave in closed-end cylinders given in Section 7.2. Here are the first two modes.

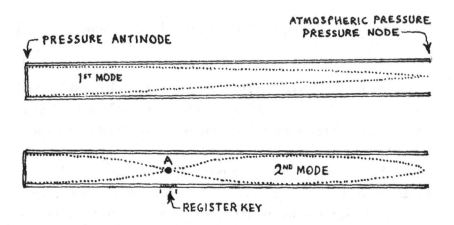

For the 2nd mode there is a pressure node (atmospheric pressure) needed both at point A and at the open end. The open end is automatically one, and an opening in the side of the cylinder at point A will tend to clamp that location at atmospheric pressure, too. The register key provides this, and it is not located halfway down length L. Instead it is two thirds of the distance back from the open end.

If you've gotten this far without looking at this answer, congratulations. You've correctly applied what you've learned about the locations of the nodes in standing waves in cylinders. Furthermore you can check the range of a B ♭ clarinet in Figure 2.2 and find that it has 26 tones, about 3 octaves. Fine, the first one-and-a-half octaves (which is also the interval between our closed-end cylinder's 1st and 2nd modes) can be played in the 1st mode using the 13 tone holes, and the remaining one-and-a-half octaves can be played in the 2nd mode with these holes. Thus, only one register key is required and it shifts the clarinet into its 2nd mode.

Answers to QUESTIONS

This is a good start, but too simple and not completely correct. Recall the warning in Section 7.2 of Chapter 7 and read on.

Look at the clarinet in the picture at the beginning of Chapter 9. The tone holes begin about one-third of the way down from the reed end.

This sketch shows the node and antinode locations for the closed-end cylinder with all its tone holes open and also all closed. As you open the tone holes you are simply playing the 1st mode of a shorter and shorter closed-end cylinder. The 2nd mode's standing wave with all the tone holes closed and open would look like this:

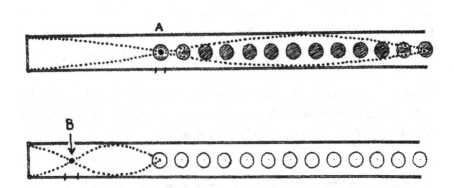

Here the trouble starts. The register key's position must be at A and then at B, and somewhere in between depending on which tone holes are open. This is a problem; where should the register key be? Can this be solved, or at least worked around? Yes, clarinets work! But, how? You'll now need some information about real clarinets.

Figure 2.2 tells you that the B♭ clarinet's range is from D_3 to G_6:

Answers to QUESTIONS

Figure 2.2 Range of B♭ Clarinet

Neglecting the sharped and flatted notes, this range included 26 notes. Here they are below, all laid out to show the three playing registers of our clarinet: Chalumeau, Clarion, and Altissimo.

$D_3\ E_3\ F_3\ G_3\ A_3\ B_3\ C_4\ D_4\ E_4\ F_4\ G_4\ A_4$ $B_4\ C_5\ D_5\ E_5\ F_5\ G_5\ A_5\ B_5$ $C_6\ D_6\ E_6\ F_6\ G_6$

        Chalumeau   Clarion   Altissimo

This strange division of the range into three registers needs an explanation. Why not just have two registers, each spanning an octave plus a MAJOR FIFTH? The Chalumeau register does this, why not both registers? The answer is that the register key would have to move from point A to point B while playing in the second register.

Note, that the Clarion register spans only one octave; the upper tone holes are always closed. This moves point B closer to point A. Below is a sketch of the 2nd mode under this condition.

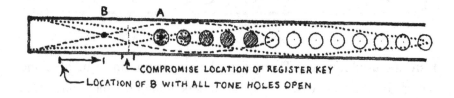

Perhaps a compromise position for the register key would now be possible between points A and B. It must be. There is still the problem of the location of a register key for the Altissimo register, especially when the QUESTION states that one register key is enough.

Answers to QUESTIONS

Now note that the frequency of the 3rd mode is five times the frequency of the 1st mode. This is an interval of two OCTAVEs plus a MAJOR THIRD, and for our clarinet makes the first note in the 3rd mode G_5. Note, too, that G_5, A_5, and B_5 are played in the Clarion register's 2nd mode; and that the highest note in the clarinet's range is G_6. So, the first three notes in the 3rd mode are not used; and this means that in the 3rd mode the three lowest tone holes are always open. The Altissimo register is from C_6 to G_6: 5 notes that use the next 5 tone holes. The upper 5 tone holes are not opened in the Altissimo register. Let's draw the standing waves for the 3rd mode for this situation, and see if a solution to our conundrum appears.

Two possible locations for a register key for the Altissimo register are indicated. They use the same kind of compromise as was picked for the location of the Clarion register's register key. One of the locations is too close to the reed; the other is at the location of a tone hole. So, the register key needed for the Altissimo register is already there, and a separate key is not needed.

Based on both the theory of standing waves in cylinders and how an actual clarinet is used, we find:

1. There are three playing registers: Chalumeau, Clarion, and Altissimo.

2. Because the upper two of these registers don't use all the tone holes it is possible to have only one register key (and use a tone hole for the other).

Answers to QUESTIONS

3. A compromise position for the Clarion mode's register key is both necessary and possible.

This is a difficult QUESTION. Our first answer was wrong because we tried to apply standing wave theory to a cylinder that had tone holes. After we found out that real clarinets had three playing registers and what their notes were, we were able to continue using standing wave theory to reach an answer. But even that answer blurred simple standing wave theory a bit by requiring a compromise position for the register key. The fact that clarinets do play as they do indicates, but does not prove, that our answer is correct.

CHAPTER 11

QUESTION 11.1: Here is one cycle of part of a standing wave shown with amplitude graphed versus the location from a node.

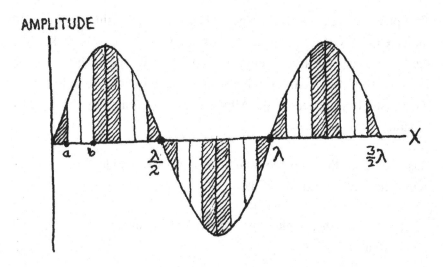

You want to show that the loudness at positions a and b are 15% and 85% of the maximum loudness which occurs at $\lambda/4$ and λ (3/4). The loudness at the edges of the other three white areas will have the same values. You can see this immediately from the graph. The equation for the graph is:

Answers to QUESTIONS

$$\text{amplitude} = A \sin(2\pi x/\lambda),$$

and you must find the amplitudes at distances a and b along the x-axis. At place a, $x = \lambda/16$, and so

$$\text{amplitude} = A \sin(2\pi/16) = A \sin(\pi/8),$$

and the loudness at place a is

$$\text{loudness} \propto (\text{amplitude})^2 = A^2(0.146) \approx 15\% \, A^2.$$

At point b, $x = \lambda (3/16)$, and so

$$\text{amplitude} = A \sin(3\pi/8),$$

and the loudness there is

$$\text{loudness} \propto \text{amplitude}^2 = A^2(0.146) \approx 15\% \, A^2$$

QUESTION 11.2: The explanation is easier. Both cases a.) and b.) in Figure 11.4 show a half wavelength spanning the distance between the ears. So both are considering the same standing wave. In case a.) the ears are at its nodes; in case b.) the listener moves so that the ears are at its antinodes: just different places on the same standing wave and thus the frequencies are the same.

Answers to QUESTIONS

QUESTION 11.3:

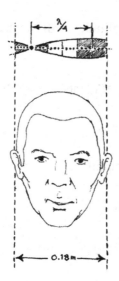

Here is a combination of Figures 11.4 and 11.5 showing the shortest wavelength that will produce nearly case c.). Each vertical stripe is $\lambda/16$ wide and there are 6 of them shown. So, $6(\lambda/16) = 0.18$ m, and $\lambda = 0.48$ m. Thus the highest troublesome frequency in this case is

$$f = (345 \text{ m/s}) / 0.48 \text{ m} = 720/\text{s}.$$

For the lowest troublesome frequency,

Answers to QUESTIONS

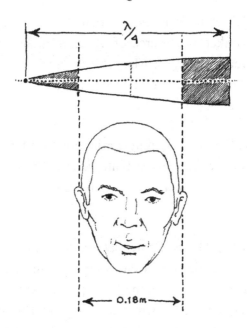

Now $\lambda \, (2/16) = 0.018$ m, and $\lambda = 1.44$ m, and $f = 240$/s.

The range of troublesome frequencies is

$$240/\text{s} \le f \le 720/\text{s}.$$

QUESTION 11.4: Recall that the width of a stripe is $\lambda/16$. One eighth of a resonant wavelength is the width of two adjacent stripes, black or white. If the wavelength were big enough so that the listener had both ears within the same two adjacent stripes, he or she would hear the sound too loudly or softly.

The total width of the two adjacent stripes would have to be at least 0.018 m, and so one-eighth the wavelength must be at least 0.018 m.

QUESTION 11.5: The listener must get both ears out of the two adjacent stripes within which the sound is too loud or soft. Even if one ear is at the edge of this bad region, the other must move a little more than 0.018 m to get out.

Answers to QUESTIONS

QUESTION 11.6: A node is the location of a high interfering with a low. A loudspeaker is a source of traveling highs and lows. If its reflected lows return to it just when it is transmitting a high, or vice versa, there will be a node there. Everything depends on how far away is the reflecting surface.

QUESTION 11.7: This seemingly straightforward QUESTION is an example of a problem without a definite answer. You will find that after a lot of calculation you will have to make a decision based on what seems reasonable.

Look again at Table 11.1 and note that after a few octaves the room's resonant frequencies include all the notes in a chromatic scale. Then it doesn't matter where you put the loud speaker; you the listener will just have to locate yourself where it sounds good. But for the first one or two octaves the room can be resonant for some of the notes and not others, and this can be bad acoustics. You want to place the loud speaker to prevent these unwanted resonances. This is done by locating the loud speaker to clamp them out.

The answer to QUESTION 11.6 showed that the loud speaker located at a possible standing wave's nodes or antinodes will cause that standing wave; just what you don't want. So, don't put the loud speaker there. But, where is "there"?

First you must find the frequencies and wavelengths of the room's lower frequency modes. Then you can find the distances from walls, ceiling, and floor of the nodes and antinodes of these modes. Finally look for positions of the loud speaker that are neither near nor at these nodes or antinodes.

Here is a list of the room's resonant modes, frequencies, and their wavelengths. These have been calculated from equations (7.2) and (2.3).

Answers to QUESTIONS

7.0 m WIDTH

MODE	f, 1/s	λ, m
1	25	14
2	50	7.0
3	74	4.6
4	99	3.5
5	123	2.8
6	148	2.3
7	173	2.0
8	197	1.8
9	222	1.6
10	246	1.4
·	·	·

10.0 m LENGTH

MODE	f, 1/s	λ, m
1	17.3	20
2	35	10
3	52	6.7
4	69	5.0
5	86	4.0
6	104	3.3
7	121	2.9
8	138	2.5
9	155	2.2
10	173	2.0
11	190	1.8
12	207	1.7
·	·	·

Answers to QUESTIONS

2.5 m HEIGHT

MODE	f, 1/s	λ, m
1	69	5.0
2	140	2.5
3	207	1.7
4	276	1.3
5	345	1.0
•	•	•

The following list presents all these frequencies in ascending order. The semitone frequencies for a chromatic scale starting at 100 Hz are also shown.

	RESONANT FREQUENCIES, 1/s	CHROMATIC SCALE FREQ, 1/s	COMMENTS
OCTAVE	17		See Figure 3.1. Even if these notes are played you probably won't hear them.
	25		
	35		
OCTAVE	50		Not enough notes here to supply the 12 needed semitones. This is the OCTAVE you don't want the loudspeaker to add resonance loudness. In fact you want to put it in place to clamp them out.
	52		
	69		
	74		
	86		
	99	100, do	
OCTAVE		106, di	Almost enough resonances in this OCTAVE to make every note a resonance: most of the resonant frequencies match up with the chromatic scale's frequencies except re.
	104	112, re	
	121	119, ri	
	123	126, mi	
	138	133, fa	
	140	141, fi	
	148	150, sol	
	155	159, si	
	173	168, la	
	190	178, li	
	197	189, ti	
	207	200, do'	

- 292 -

Answers to QUESTIONS

You can see why no more room resonant frequencies were calculated. There will be plenty of match ups between the needed and available resonances for the higher octaves.

Next calculate the distances of the nodes and antinodes from a surface. Use the figure below to remind you where they generally are, and plug in the wavelengths already found above for the middle octave.

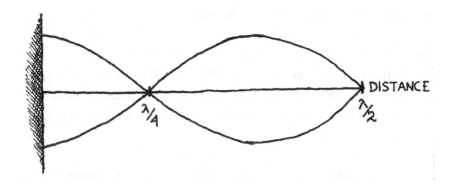

Then show the distances on a chart:

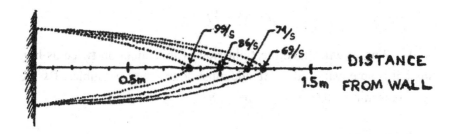

So, put the loud speaker about 0.5 m from a surface. This is away from nodes and antinodes, and a reasonably close distance from the surface. You probably would have put the loud speaker there without doing any calculations. In any case you should adjust its position to achieve a sound you like.

QUESTION 11.8: The method for answering this QUESTION is the same as for the previous one except that you must use the formula

Answers to QUESTIONS

for a cavity open at one end, equation (7.1), to calculate some of the frequencies. The numbers will be the same for the other two dimensions; they are still closed at both ends.

APPENDIX A

QUESTION A.1: It's impossible to count the air molecules; there are too many. But, this QUESTION asks for a way to change this number by a known percentage. This can be done.

Take a flask of air at any pressure and connect it to an identical flask containing a vacuum. After a while the original air molecules equally fill both flasks, and each has half. You have decreased the number of air molecules in the original flask by 50%. Continue doing this to get 25%, etc.

If you want some other percentage change use unequal volume flasks.

QUESTION A.2: The hint gets you to the equation

$$p = (kT/V)\,N\,.$$

So doubling N will make the right side of this equation twice as big. Since it is an equation the left side must double too: p doubles. By the same reasoning halving N will halve p.

For the syringe,

$$pV = \text{constant}.$$

Halving V, by pushing down the plunger, causes p to increase. The product pV must stay constant; so halving V will cause p to double. Similarly, making the volume one-third its original size will cause the pressure to triple.

Answers to QUESTIONS

QUESTION A.3: The "constant" in

$$T = (\text{constant}) \, (v^2)_{ave}$$

is not the same one you used in the last QUESTION. You should find that

$$p = (N\,k/V)(\text{constant})\,(v^2)_{ave} = (\text{another constant})\,(v^2)_{ave}.$$

Now doubling p will double the value of $(v^2)_{ave}$; and thus p is proportional to $(v^2)_{ave}$. This is just what the second experimental result says.

QUESTION A.6: You are adding molecules to a more-or-less fixed volume. The temperature stays about the same. So from

$$pV = kNT,$$

$$p = (k\,T/V)\,N = (\text{constant})\,N, \text{ with yet another constant.}$$

Increasing N will increase p. The pressure increase is predicted by the Ideal Gas Law.

None of this equation manipulation actually tells you what's happening; it only predicts that adding more air molecules will increase the pressure.

The mechanical model of a gas shown in DEMONSTRATION XXIV does give you something else to talk about. It showed that small spheres, representing air molecules, beating against disc D2 would exert an upward force on it. More spheres caused more beats and D2 would rise unless you added weight to keep the volume between the discs the same. Thus adding spheres has increased the "pressure" against the under side of D2. This "pressure" increase could also happen if each sphere hit harder, but that would require the spheres to have bigger speeds, and according to the Kinetic Theory of Gases this would need a temperature change, which is not happening in this QUESTION's

Answers to QUESTIONS

situation. On the other hand, driving on a hot day will increase your tire pressure.

QUESTION A.7: Everything can be calculated if you remember the definition of density: the number of kilograms of some thing in a cubic meter of its volume. For air at 20 °C and 1 atmosphere pressure this is

$$m\,N/V,$$

where V is one cubic meter, m is the mass of one air molecule, and N is the number of air molecules in this cubic meter.

First let's calculate the number of air molecules in a cubic meter of air: N/V. From the Ideal Gas Law,

$$pV = kNT,$$

we can show that

$$N/V = p/kT = 1.0 \times 10^5 \text{ Pa} / ((1.4 \times 10^{-23} \text{ kg m}^2/\text{s}^2 \text{ K})(293 \text{ K})).$$

The units in the above formula are bizarre, but once you are told that 1 Pa = 1 kg / s² m, a strange but correct way to specify a pressure, you can write

$$N/V = p/kT$$

$$= (1.0 \times 10^5 \text{ kg/s}^2 \text{ m}) / ((1.4 \times 10^{-23} \text{ kg m}^2/\text{s}^2 \text{ K})(293\text{K}))$$

$$= 2.4 \times 10^{25}/\text{m}^3.$$

This is a huge number. It is said that if the gasp of air that Julius Caesar exhaled during "Et tu, Brute?" was evenly distributed throughout the earth's atmosphere we would inhale about a thousand of these air molecules in each breath.

Answers to QUESTIONS

You have read that the density of air is 1.2 kg/m² = m (2.4 x 10^{25}/m³), and so the mass of one air molecule is

$$m = (1.2 \text{ kg/m}^2) / (2.4 \times 10^{25}/\text{m}^3) = 5.0 \times 10^{-26} \text{ kg}.$$

To determine the average distance between air molecules, let's suppose you have a cubic meter of them in a cube

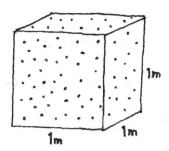

The molecules are about equal distances apart. Call this distance, d. So d^3 is the volume occupied by each molecule. This volume multiplied by the number of molecules must be one cubic meter, or

$$d^3 (2.4 \times 10^{25}/\text{m}^3) = 1 \text{ m}^3, \text{ and } d = 3.5 \times 10^{-9} \text{ m}.$$

They are close together, and this result also indicates that an air molecule must be smaller than this.

To find the average speed of an air molecule and the approximate number of collisions per second we must use equation (A.2):

$$(1/2) \, m \, (v^2)_{ave} = (3/2) \, k/T.$$

Everything except $(v^2)_{ave}$ is known and so

$$(v^2)_{ave} = 3kT / m$$

$$= 3 \, (1.4 \times 10^{-23} \text{ kg m}^2/\text{s}^2 \text{ K}) \, (293 \text{ K}) / 5.0 \times 10^{-26} \text{ kg}$$

Answers to QUESTIONS

$= 2.46 \times 10^5$ m2/s^2, and the average speed is

$$\sqrt{(v^2)_{ave}} = 500 \text{ m/s}.$$

The number of collisions per second a single air molecule makes is the inverse of the time between collisions.

From

$$\text{speed} = \text{distance/time},$$

the time to go 3.5×10^{-9} m at a speed of 500 m/s is

$$\text{time} = 3.5 \times 10^{-9} \text{ m} / 500 \text{ m/s} = 7.0 \times 10^{-12} \text{ s},$$

and the number of collisions per second is 1.4×10^{11}/s.

QUESTION A.8: Do the algebra necessary to combine

$$(1/2) \, m \, (v^2)_{ave} = (3/2) \, k / T \text{ and } pV = kNT \text{ to produce,}$$

$$(1/2) \, m \, (v^2)_{ave} = (3/2) \, pV / N.$$

Then isolate p on the left side of the equal sign to get:

$$p = (1/3) \, (mN / V) \, (v^2)_{ave},$$

and in our case, because the number of molecules and the volume are fixed,

$$p = (\text{constant}) \, (v^2)_{ave}.$$

Doubling $(v^2)_{ave}$ will double p.

QUESTION A.9: The surface area of a sphere having radius r is $4\pi r^2$. If this surface represents an expanding high, the excess number of

Answers to QUESTIONS

molecules is at this surface. So, the number of these molecules per square meter is:

a fixed number of molecules / $4\pi r^2$,

and as the sphere gets bigger this ratio gets smaller. It varies as $1/r^2$. This is called the inverse square law. The molecules are neither created nor destroyed, just spread over a larger and larger surface.

The last sketch in Chapter 2 shows puffs leaving the loud speaker and entering the mike. It looks like the puffs didn't lessen during this trip, and this is not so. The puffs are just small parts of the expanding sound wave. Each puff, according to the inverse square law, becomes a little bit less high as it travels. This refinement was not needed in Chapter 2.

QUESTION A.10: Figure A.1 shows that the strongest diffraction would happen when the obstacle and the sound wavelength are the same size. Persons are about 0.6 m wide, and they sit about 0.6 m apart. So,

$f = v / \lambda = $ (345 m/s) / 0.6 m = 575/s .

Figure 2.2 indicates that 575/s is near D_5, a common musical tone.

QUESTION A.11: Air by itself is a transmitter, but nobody calls it that. A horn's bell reflects part of the sound wave back into the horn and transmits the rest. This is partial transmission, just as it is partial reflection

An absorber that does not absorb all the sound hitting it either reflects or transmits the rest.

Using one of the phrases "partial reflector" or "partial transmitter" directs your thoughts to the sound inside or outside the cavity. This seems reasonable if one of these is what you want to emphasize. You

Answers to QUESTIONS

can decide for yourself; but generally "transmitter" is seldom used this way.

QUESTION A.12: Look again at Section 8.2.1 in Chapter 8, and find the sentence "There is a rule of reflection that is applicable at both the open and closed end of a cavity: the sound will be at least partially reflected at places along the cavity where the pressure changes abruptly." The part of this following the colon can be rewritten to read: " ...end of the cavity: the sound will be at least partially reflected at places along the cavity where the impedance abruptly changes." This happens at both the open and closed ends of a cavity, but also at any place within the cavity where there is an abrupt impedance mismatch.

Of course, you shouldn't use the word "impedance" until it has been described as in Section A.4.2.4.

APPENDIX B

QUESTION B.1: The time for the direct sound to reach L1 is

$$t = d / v = 5.0 \text{ m} / (345 \text{ m/s}) = 0.014 \text{ s}.$$

The sound from speaker a. should arrive between 0.035 s and 0.060 s later: between 0.049 s and 0.074 s after the sound leaves the musicians.

Answers to QUESTIONS

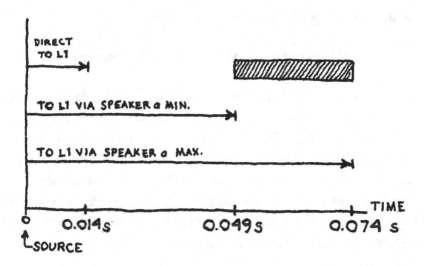

Speaker A's sound should arrive at L1 some time within the shaded time region in the above diagram. Speaker A's sound has the following delays:

source to mike + sampling interval + computing time + speaker A to L1. Therefore,

$t = (2.0 \text{ m} / (345 \text{ m/s}) = 0.006 \text{ s}) + 0.020 \text{ s}$

$+ ? + (7.0 \text{ m} / (345 \text{ m/s})$

$= 0.021 \text{ s})$.

Neglect the time it takes the electric signals to travel through wires; their speed is the speed of light.

So,

$0.049 \text{ s} \leq$ computing time $+ 0.047 \text{ s} \leq 0.074 \text{ s}$, or

$0.002 \text{ s} \leq$ computing time $\leq 0.027 \text{ s}$, for listener L1.

Answers to QUESTIONS

Do the same analysis for listener L2 and find,

0.031 s ≤ computing time ≤ 0.056 s, for listener L2.

The necessary computing times for listeners L1 and L2 do not overlap, and strictly speaking you will need two computers generating FFTs and additional time delays. Do you want to use two computers doing this? Or do you think that 0.027 s is not that much different than 0.031 s, and perhaps one FFT and time delay totaling 0.029 s would be OK for both listeners? If so, you can use as much of the 0.029 s as you want for calculating the FFT. In 0.029 s a 2 GHz computer clock will tick 58 million times, and this seems enough to do the FFT.

QUESTION B.2: Most of Appendix B describes computer-generated corrections to a hall's acoustics. These techniques can change the sound's delay time, loudness, and *RT*. This QUESTION B.2 only asks you to deal with problems of loudness and *RT*.

Loudness is due to all the sound you hear, both the **direct sound** and any later **indirect sound**. The direct sound must arrive first to the listener. This, as you've read in Section 11.1.2.3 of Chapter 11, will give him or her the impression that the sound is coming from the musicians. Direct sound now includes both the sound directly from the musicians and from loud speakers placed near the musicians and directed at the audience. Section 11.1.2.3 also says that all this sound must reach the listener within 0.035 s after its first arrival. All the indirect sound must reach the listener within 0.060 s after the first arrival.

The *RT* depends on the original loudness of the sound and what fraction of it has been lost every time it passes the listener due to reflections, diffractions, and absorptions. The direct sound has no *RT* until it reflects from some object in the hall and becomes indirect sound. Indirect sound can also be generated by C-CAE, and have its own loudness, *RT* and delay time. It is played through the indirect sound loud speakers located throughout the auditorium. They are placed beside the listeners

Answers to QUESTIONS

so that their sound will not be heard as direct sound. Microphones near the musicians can be the source of both direct and indirect sound. Microphones placed throughout the hall can monitor the sound there and tell the C-CAE that sound corrections might be needed.

Figures B.4 and B.4.A show how RT can be lengthened with and without increasing the loudness of the direct sound.

The chart below shows possibilities for changing RT. Both C-CAE and physical changes to the auditorium are indicated. The dashed paths remind you that a change in the direct sound's loudness will cause a change in some of the indirect sound.

Answers to QUESTIONS

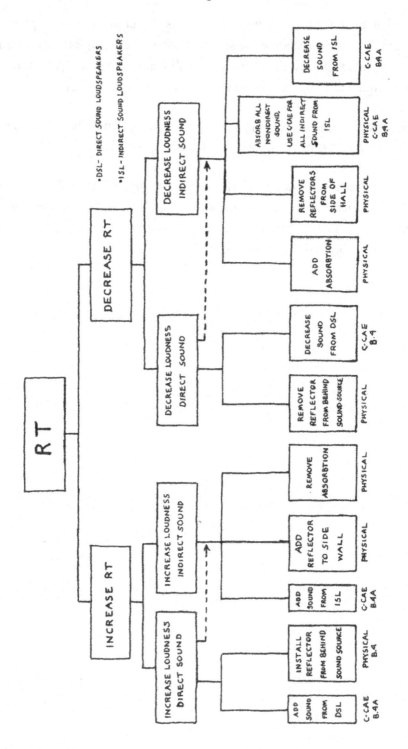

Answers to QUESTIONS

Now you can recommend changes to correct the various defects described in this QUESTION by picking and choosing *RT* and loudness changes from those available in the above chart.

Loudness OK (direct sound OK)

RT too long

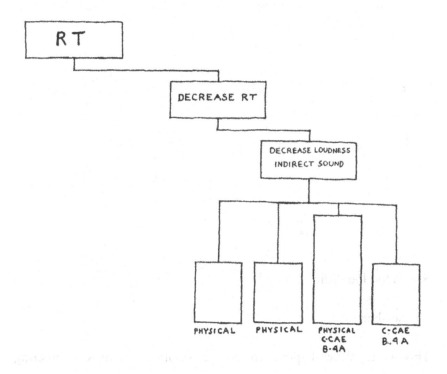

Loudness OK (direct sound OK)

RT too short

Answers to QUESTIONS

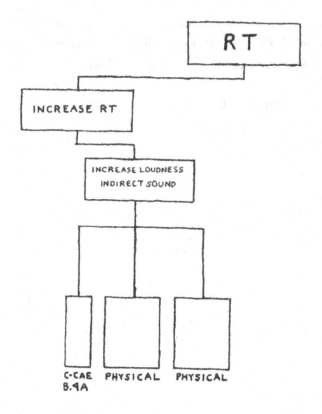

Sound too loud (direct sound)

RT too short

This is a difficult problem whose solution requires conflicting changes.

Answers to QUESTIONS

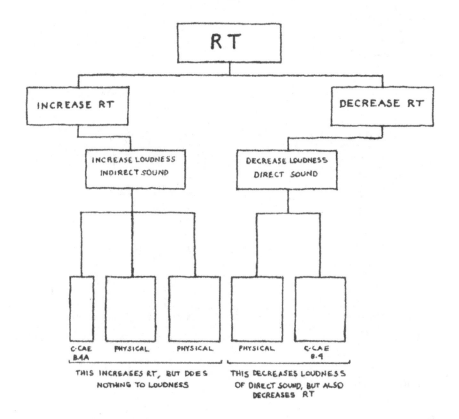

So, you must increase *RT* by computer techniques more than enough to just produce the wanted *RT*. This increase must also make up for the *RT* loss due to the reduced loudness of the direct sound.

Sound too loud (direct sound)

RT too long

Answers to QUESTIONS

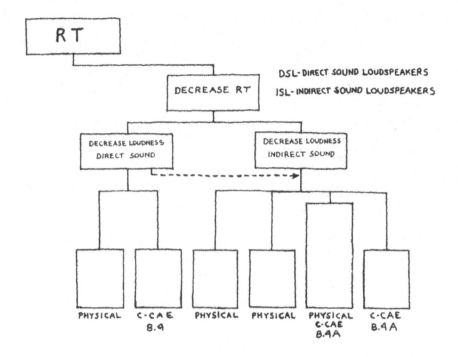

Sound not loud enough

RT too short

Answers to QUESTIONS

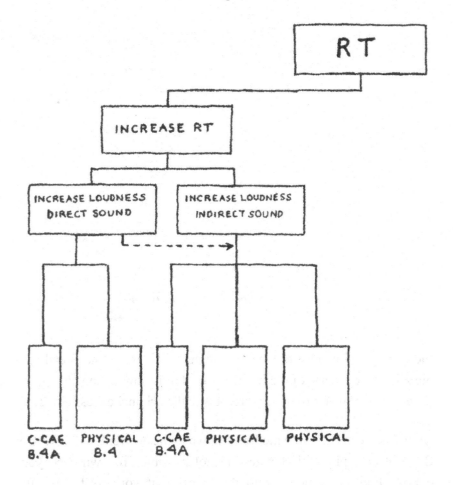

Sound not loud enough

RT too long

Answers to QUESTIONS

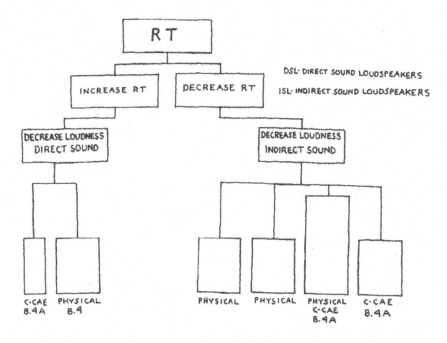

Increasing the loudness of the direct sound will increase the already too big *RT*. The decrease of the indirect sound must lower the value of the *RT* enough to overcome this increase and the original excess of *RT*.

QUESTION B.3: Use the methods shown in the answer to QUESTION B.2, but now physical changes are not possible. You will find your options limited, sometimes to the extent that you can't solve the problem.

QUESTION B.4: A lack of *RT* may be making the sound "dry." Let's make a block diagram that shows how to handle this problem. You can assume that the lack of sufficient *RT* has been identified. Also you know both the present and WANTED *RT*. In a recording studio these can be determined by replays of the recorded sound. For live performances it must be already known; perhaps the musicians can spot problems and make choices based on rehearsals.

For a general lack of *RT* you will add loudness to all the frequencies in an attack. The best way to do this is without the use of C-CAE: just put

Answers to QUESTIONS

a microphone near the sound source, connect it to an audio amplifier, and play a louder sound through direct sound loud speakers (DSL). This is the easiest way. However, if no more direct sound loudness can be tolerated, you must add loudness to the indirect sound using the method shown in Figure B.4.A. The following block diagram shows a way to do accomplish this.

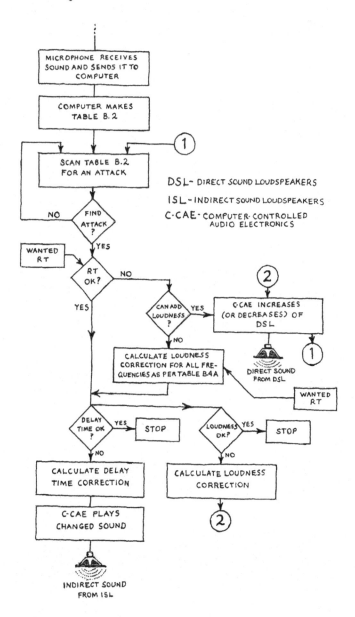

Notes

Chapter 1

1. This is Band 1 of the recordings that you can hear on the web page, www.thestructureofmusicalsound.com. See Recordings for a listing of these.

Chapter 3

2. See the book, *The Science of Sound*, Thomas D. Rossing, 1982, Addison-Wesley Publishing Co. Chapter 8, and in particular, Section 8.10. discusses the relationship between phase and timbre. It appears that most of the experimental investigations of this relationship have been done under rather tightly controlled laboratory conditions that might not obtain in everyday music.

3. This definition is taken from *The Penguin Dictionary of Physics, 2nd ed.*, Valerie Illingworth ed., Penguin Books, Inc.

Chapter 5

4. Descriptions of these instruments can be found in, *Musical Instruments of the World*, the Diagram Group, 1976, Facts on File, Inc.

Notes

5. See the book, *The Science of Sound*, Thomas D. Rossing, 1982, Addison-Wesley Publishing Co., Chapter 14.

6. Reported by Carleen Maley Hutchins, *ThePhysics of Violins*, Nov., 1962, Scientific American

Chapter 7

7. The *Visual Display of Quantitative Information*, Edward R. Tufte, 2001, Graphics Press LLC, Cheshire, Connecticut, USA

Chapter 8

8. Arthur H. Benade's article "The Physics of Brasses" in the July 1973 issue of Scientific American magazine presents a thorough explanation of the bell's operation, but with accompanying mathematics.

Chapter 10

9. Herman L. F. Helmholtz (1821-1894), physicist and physiologist. Helmholtz' book *On the Sensations of Tone* is listed in the Additional and Extended Readings. Helmholtz first stated the Law of Conservation of Energy.

Chapter 11

10. The recent history of acoustical design is a story of trial and error, success and failure. Scientific and engineering principles, when heeded, do seem to contribute to better results. *The Science of Musical Sound*, John R. Pierce, 1983, Scientific American Books, gives a vivid account of a recent failure. Tom Manoff's article in the 31 March 1991 New York Times (reprinted in this book) and James Glanz' article in the 18 April 2000 New York Times describe several successes.

Notes

11. Manfred Schroeder (1926-), Professor of Physics and Director of the Drittes Physikalisches Institut at the University of Gottingen. See his article in the *Journal of the Audio Engineering Society*, 32 (4), pp. 194-203, April 1984. Professor Schroeder's lively prose makes this article pleasant reading, and accessible up to the place where some advanced mathematics is used.

12. These recipes are adapted from *Music, Speech, High Fidelity, 2nd ed.*, William J. Strong and George R. Plitnik, 1983, SOUNDPRINT; and *The Science of Sound*, Thomas D. Rossing, 1992, Addison-Wesley Publishing Company.

13. See note 11, above.

Appendix B

14. J. W. Cooley, J. W. Tukey, "An Algorithm for the Machine Calculation of Complex Fourier Series." *Math. Comp.* 19 (90), pp. 297-301, April 1965

Appendix C

15. The phrases "Physical Qualities" and "Physical Objects," and the notation for these things are from *The Penguin Dictionary of Physics, 2nd ed.*, Valerie Illingworth ed., Penguin Books, Inc.

Glossary of Technical Terms

Some of the words explained in this glossary have both scientific and musical meanings, but only the scientific ones are given. And even these are restricted to musical sound.

A complete collection of musical meanings can be found, for example, in *Music Theory Dictionary: The Language of the Mechanics of Music*; William F. Lee, editor; published by Charles Hansen Educational Music and Books, 1966.

absolute zero temperature. The lowest temperature possible. The temperature at which air molecules would cease to move. It is -273 °C or 0 K. on the Celsius or Kelvin temperature scales.

absorption. The process of sound being lost due it changing into something we cannot hear.

active acoustics. Room acoustics that can be changed rather quickly by physical and/or electronic means. Computer-Controlled Audio Electronics (C-CAE) is an example of a method of achieving this.

algebraic addition. The "addition" of several values taking into account that they can be positive or negative. So, algebraic addition can be addition, subtraction, or both in some cases. Examples: $5 + 3 = 8$, $-2 - 3 = -5$, $5 + 3 - 2 = 6$.

Glossary

amplitude. The maximum amount that a Physical Quantity can differ from its normal value.

analogue. A word that describes a quantity that changes smoothly from one value to another one. Analogue ≠ digital.

analogue-to-digital conversion, (ADC). A method that changes the smoothly changing values of a quantity into a progression of small step-sized changes. Also ADC is the name of the result of making an analogue-to-digital-conversion.

antinode. The location along a standing wave where its displacement can become its amplitude. Antinode ≠ node.

audio oscillator. An instrument that generates electrical oscillations whose frequencies lie within the range of those of audible sound. An audio oscillator connected to a loud speaker will produce sound.

bandpass filter. An instrument that blocks the passage of all frequencies except a single one, or a narrow range (band) of frequencies.

bar. A unit of pressure. One bar is normal atmospheric pressure.

beat. The rather slow oscillation of the loudness of a sound.

bell jar. A large glass bottle which can be connected to a vacuum pump and have the air inside evacuated. It has a sealable door so that things can be put inside, and electric wires can enter into it.

block diagram (flow chart). A simple chart which is read from top to bottom, and which shows how the particular steps occur during a complicated process.

boundary condition. The rule that must be obeyed by sound when it reaches the boundary between one type of medium and another.

Glossary

cavity-controlled oscillator. A prototype musical wind instrument consisting of a cavity, a reed, and an air supply. The shape of the cavity determines the frequency sound it can play. The interaction between the sound wave in the cavity, the reed and the air supply sustains the sound.

Chladni figures, plates. Chladni figures are the locations of where the sand ends up on a flat plate vibrating at one of its modes. These are the nodes of the standing waves. The sand is bounced away from the antinodes. Chladni plates are the flat plates on which Chladni figures are formed.

chromatic scale. The 13 tones comprising an octave interval: do, di, re, ri, mi, fa, fi, sol, si, la, li, ti, do'. The tones are a semitone apart, and each tone's frequency is about 6% bigger than the preceding tone's. Often just called the scale. See scale.

clamping. Any method that forces a standing wave to have a node at the positions of the clamp. All the possible standing waves that could not have nodes at this position are prohibited: clamped out.

complex wave. A sound wave that contains several frequencies.

complex vibration. The oscillation with several frequencies of a Physical Object.

combination frequency. An additional frequency created by a transducer or amplifier that does not pass or amplify a sound faithfully.

Computer-Controlled Audio Electronics (C-CAE). A sound generating system used to change a room's acoustics.

cycle. The reoccurring part of a vibration or wave.

Glossary

decibel (dB). A non-metric unit of sound pressure. It compares a sound wave's pressure fluctuations to the pressure fluctuations of a just audible sound.

delay time. The time interval between when the musical sound arrives at the listener directly from the musician(s), and when it arrives after a reflection or some other delay. This is an acoustical property of a room or hall.

digital. A word that describes something that changes value a step at a time. Digital $\neq$ analogue.

digital-to-analogue conversion (DAC). A method that converts step-by-step changing values into a smoothly, or continuously changing one. DAC must create values in between the steps, and these might not be right. DAC $\neq$ ADC.

diffraction. A process in which sound waves change direction, but not frequency, when they encounter obstacles. Diffraction, unlike reflection, describes how sound waves continue "behind" the obstacles.

direct sound. Musical sound that reaches the listener on a straight line path from the direction of the musician(s).

displacement.. The amount of change of the value of a Physical Quantity.

early field system. The microphones, loud speakers, and C-CAE that artificially produce the first "reflected sound" the listener hears.

ensemble. The characteristics of a room's acoustics that allow the musicians to hear each other.

Glossary

envelope. Two lines on an amplitude vs. distance graph which represent the maximum and the minimum values of an amplitude at that distance, at any time.

Fast Fourier Transform (FFT). A computer program that speeds up the time it takes to produce a Fourier transform. Also the resulting Fourier Transform produced by the FFT.

filter. A device that separates out, and passes on, previously chosen characteristics of musical sound, such as certain frequencies or loudnesses.

fipple. The reed of an air reed instrument. It directs the air puffs to enter, or not enter, the cavity.

First Law of Thermodynamics (Conservation of Energy). In an isolated region, the amount of energy is fixed. The energy can change form, but not amount.

flat. The name of the symbol, ♭ , written next to a note. This decreases the pitch of the note by one semitone, or equally the frequency of the note by about 6%. See sharp.

flow chart: See block diagram.

Fourier analysis. A process that determines the frequencies and their amplitudes of all the single frequency waves that have created any complex wave.

Fourier transform. The table of the frequencies and amplitudes of the single frequency waves resulting from making a Fourier analysis.

frequency. The number of times per second something is repeated. It is also the number of cycles per second.

Glossary

fundamental. The name of the lowest frequency mode of vibration. Also called the 1st mode.

gauge pressure. The amount of pressure difference from atmospheric pressure.

hard-wired. An adjective describing electronic apparatus that can do only one thing, and in the case of computers do not have to step through their jobs according to the ticking of the computer's internal clock.

harmonic oscillator. A vibrator whose allowable frequencies are integer multiples (also called harmonics) of its fundamental frequency.

harmonics. The frequencies of the higher modes of a harmonic oscillator. The frequency of the 1st, or fundamental, mode is called the fundamental. The frequencies of the higher modes are called the harmonics, and numbered 1st, 2nd, etc. Overtones are also harmonics if their frequencies are integer multiples of the frequency of the fundamental mode.

Helmholtz resonator. A cavity with several parts. Some of them act like a spring in a spring-bob oscillator and the others act like the bob.

Hertz (Hz). A name for the unit of frequency. 1 Hz = 1 cycle/second = 1/s.

Ideal Gas Law. The rule which shows the relationship between a gas' pressure, volume, temperature, and how many gas molecules are present: $pV = kNT$. The Ideal Gas Law, when applied to air, assumes that air molecules are simple hard spheres.

indirect sound. Musical sound that does not reach the listener on a line-of-sight path from the musicians. Reflected sound is indirect sound; so is the sound from any loud speakers beside the listener.

Glossary

initial loudness. The loudness of the tone the listener hears first.

integer. A whole number.

integer multiple. An integer multiple of a Physical Quantity is the value you get by multiplying that Physical Quantity by a whole number. The values of the possible resonant frequencies of a string, cone, or cylinder are integer multiples of the frequencies of their fundamental modes of vibration. The values of the vibration of a harmonic oscillator are integer multiples of the frequency of its fundamental mode of vibration.

intensity. The rate of flow of sound energy through a standard cross sectional area. Intensity is measured by technical instruments and not by the ear. Intensity has a completely different musical definition. Intensity ≈ loudness.

interface. A boundary between two media or between two parts of the same medium in different conditions.

inverse-square law. A rule that explains why sound in open air gets less loud as you move away from its source.

Kinetic Theory of Gases. A proposal that gas is a collection of hard spheres, called molecules, which fly around colliding with each other and with the walls of their container. These collisions create what is perceived as the gas' pressure.

loudness. A subjective name for a property of sound that we hear. What you would call a music's loudness depends on how good your hearing is. It also depends on the sound itself: increasing the concentrations of air molecules in the sound generally produces more loudness.

macroscopic property. A property of something that is apparent in the everyday size of that thing. Macroscopic property ≠ microscopic property.

Glossary

mass. The property of an object that makes it difficult to change its motion. This is not the only definition of mass.

mathematical symbols.

$\pm$; "plus or minus the value of"

$\geq$; "has a value equal or greater than"

$\leq$; "has a value equal or less then"

$\approx$; "has a value approximately the same as"

$\neq$; "is not or does not have the same value as"

$=$; "is or has the same value as"

$\propto$; "is proportional to "

microscopic property. A property of something that is apparent only when it is examined in great detail or on a very small scale. Microscopic property $\neq$ macroscopic property.

mode. The general name of an object's permitted type of vibration. The 1st mode of an object's vibration is also called its fundamental mode or just its fundamental. Also a musical term specifying the number of semitone intervals between notes in a scale.

molecule. A molecule is a combination of atoms held together as one body by their electric or magnetic forces. As far as musical sound is concerned air molecules are just small hard spheres. The details of their construction are not needed.

near wall effect. Sound hits a wall and some is reflected. The superposition of the incident and reflected sound causes nodes and antinodes close to the wall. This is the near wall effect. Unlike standing

Glossary

waves in strings and cavities, near wall effects can happen for any incident sound's frequency.

node. The location along a standing wave where the vibrating medium does not vibrate. Parts of a vibrating string are at rest at its nodes, and air remains at atmospheric pressure at the nodes of its standing waves. Node ≠ antinode.

nonlinear. An adjective that means that the ratio of cause to effect is not constant. Your ear does not have the same sensitivity to all frequency tones. It is a nonlinear device. Any sound system that does not amplify all frequencies the same is nonlinear.

note. The written symbol on a musical staff for a tone with a specified frequency.

octave. The musical interval between two tones, one with twice the frequency of the other.

oscilloscope, scope. An electronic instrument that displays graphs of the amount of an electric signal vs. time. If the electric signal is the output of a microphone, the scope will present a graph of the sound level changing with time.

overtone. A higher pitch tone generated simultaneously with the fundamental mode's tone. An overtone may or may not be a harmonic. Overtone is another name for a partial.

parallel computing. A timesaving method of computing. Several computers simultaneously perform different subparts of a computer program, and share results as needed. Compare this to having a single computer do all the parts, but one after another.

partial. See overtone.

Glossary

partial vacuum. Gas with a pressure less than one atmosphere. See vacuum.

permutation. One of the possible ways to arrange a group of things.

period. The number of seconds that occur during a cycle. The number of seconds it takes one cycle of a wave to pass some point.

phase. A word that describes if two waves are synchronized or not. If they both begin a cycle at the same time and are both increasing or decreasing together, they are "in phase." If they both begin a cycle together but one is increasing and the other decreasing, they are "completely out of phase." Any other combination of starting times and increasing or decreasing is called "being out of phase."

Physical Object. A thing that you can touch, see, or hear (and possibly smell or taste). It can have size or other characteristics, but their values do not have a number with metric units. The symbols for Physical Objects are not written in italics.

Physical Quantity. A thing whose value is a number with metric units. The symbols for Physical Quantities are written in italics.

pitch. The frequency of a tone you hear.

pressure. A macroscopic property of a gas caused by its molecules beating against some surface or each other. For musical sound the molecules are the air and the pressure depends almost wholly on the local excess or deficient concentrations of these molecules as the sound wave passes through the air. Technically, the amount of force against some area divided by the area is the pressure at this area.

reflection. When sound arrives at a boundary between two environments at least part of it bounces back. This bounce is called reflection. The reflected part has the same frequencies as the incident sound. You

Glossary

hear it as the same music, but perhaps less loud depending on what fraction of the original sound is bounced back (reflected). This fraction is complicated to describe. It depends on the materials that make up the boundary and the size of the boundary. Generally, hard surfaces and boundaries having dimensions larger than the wavelengths of the sound reflect a bigger fraction of the sound. However, because the range of wavelengths of audible sound is from about 0.02 m to 7 m, reflecting surfaces may not reflect all the frequencies equally well.

register key, speaker hole. A hole is the side of woodwinds that, when opened, causes the woodwind to play in a higher mode. The register key is placed along the cavity where the higher mode would have a node. The open register key thus tends to clamp the cavity there at atmospheric pressure to produce this needed node.

reverberation field system. A collection of microphones, audio electronics, and loudspeakers that continues to broadcast sound into the auditorium. This increases the perceived reverberation time.

reverberation time, *RT*. The number of seconds it takes multiply reflected sound to become inaudible. This is an acoustical property of a room or auditorium. Reverberation time can have a different value when measured by instrumentation. Then it is the number of seconds it takes the sound's loudness to reduce to a standard fraction of its original loudness. This is also an acoustical property of a room or auditorium.

resonance. When a vibrator is oscillating at, and sustaining, a particular frequency of vibration, it is in a resonance. Musical instruments are made so that all the pitches they are designed to play are their resonances.

resonant frequency. The frequency of a mode of vibration.

scale. The 8 notes in an octave: do, re, mi, fa, sol, la, ti, do'. See chromatic scale.

Glossary

Second Law of Thermodynamics. A rule that states, among other things, that noise by itself will not organize itself into musical sound. This Law provides a philosophical basis for this book's definition of musical sound.

semitone. The pitch interval between the notes in a chromatic scale. A pitch increases, or decreases, by a semitone when its frequency is increased, or decreased, by about 6% (5.946%).

sharp. The name of the symbol, ♯, written next to a note. This increases the frequency of the note by about 6%, a semitone. See flat.

side hole, tone hole. An open-able hole in the side of a wind instrument. Side holes are opened or closed to cause the instrument to play the various notes in a chromatic scale.

soundboard. A flat plate or membrane connected to the strings of a musical instrument. Its purpose is to transmit the vibrations of the strings to a larger surface whose vibrations will move more air, and cause louder sound.

soundbox. Usually an irregularly shaped box, whose top and bottom are soundboards.

speaker hole. See register key.

spring-bob oscillator. A coiled spring suspended from one end and with a weight, a bob, attached to the other.

standing wave. The wave motion caused by two equal frequency waves traveling through each other in opposite directions. The two waves are often one a reflection of the other. The standing wave has nodes and antinodes, and does not appear to be traveling. Standing waves cause resonances.

Glossary

superposition. The technique of adding, or subtracting, (algebraic addition) the displacements of two intersecting waves to determine the resulting displacement at the point of intersection.

Applying the technique of superposition to the two intersecting traveling waves produces the information for making the graph of the displacement of a complex wave. Mother Nature automatically does a superposition of intersecting sound waves and creates the resulting complex wave. If the amplitudes of the two intersecting waves are too big, superposition doesn't work and the amplitude of the complex wave will not be able to be big enough. This is a nonlinear situation.

tempering. A name for some method of determining the relative frequencies of the notes in a scale.

timbre. A general name for the sound made by several tones heard together.

tone. The sound of the played note. Tones have pitch, loudness, timbre, and duration.

tone hole. See side hole.

transducer. A device that changes sound into a corresponding electrical signal, or vice-versa. For example: a microphone or a loudspeaker.

units. The basic amounts in which the values of Physical Quantities are stated.

under balcony system. The system of microphones, audio electronics, and loud speakers that improves the acoustics under an auditorium's balcony.

vacuum. A volume containing no gas molecules. See partial vacuum.

Glossary

vacuum pump. A machine that removes gas molecules from some volume, such as a bell jar.

wavefront. The part of a traveling wave that arrives first.

wavelength. The length of a cycle of a wave. How far a wave travels in a time interval equal to its period.

Additional and Extended Readings

The first five books provide alternate or extended explanations of the topics in *The Structure of Musical Sound* at about the same level. They can be read, referred to, and enjoyed by the readers of *TSMS*.

The Physics of Music, introduction by Carleen Maley Hutchins, 1978, Scientific American Books, W. H. Freeman and Company. Reprints of articles from Scientific American. Includes the singing voice, piano, woodwinds, brasses, violins and bowed strings, and architectural acoustics

Horns, Strings & Harmony, Arthur H. Benade, 1960, Anchor Books, Doubleday & Company, Inc. This book was one of the results of a national effort in the 1950s and 60s in the United States of America to improve science teaching and learning, especially in secondary schools. Benade does not begin with the usual mathematical preambles. In fact, his book contains almost no mathematics. It concentrates on the ideas of musical sound: what it is and how it is produced. This was a groundbreaking method then and, to some extent, is still one today. The combination of good science and lively writing has been hard to beat. *TSMS* has been strongly influenced by Benade's example.

The Science of Musical Sound, John R. Pierce, 1983, Scientific American Books, W. H. Freeman and Company. A unique book. Seemingly written for the musician, it is also quite technical. Pierce's interest is in computer-made music, and his book contains very little about traditional

Additional and Extended Readings

musical instruments. Nevertheless, he speaks directly to the scientific and engineering properties of music and the mechanics of its creation.

Musical Instruments of the World, An Illustrated Encyclopedia, The Design Group, 1976, Facts on File, Inc. More than 4000 original drawings. Mostly pictures of musical instruments and how they are played. Just enough text to accompany the pictures.

The Science of Sound, Thomas D. Rossing, 1982, Addison-Wesley Publishing Company, Inc. As its title states this book covers more than musical sound. It is an excellent reference.

The following books are more advanced.

Fundamentals of Musical Acoustics, Arthur H. Benade, 1976, Oxford University Press, New York, NY. A comprehensive scientific description of musical sound, musical instruments, and acoustics. Benade worked with musicians and this book contains a lot of practical information.

Musical Acoustics, Selected Reprints, Thomas D. Rossing, ed., 1988, American Association of Physics Teachers, College Park, MD. In addition to the reprinted articles this book contains a complete "road map" to the scientific literature about musical sound.

On the Sensations of Tone as a Physiological Basis for the Theory of Music, The Second English Edition translated by Alexander J. Ellis, Herman L. F. Helmholtz, 1954, Dover Publications, Inc. New York, NY. Generally readable, except for the mathematical Appendices. This is the earliest book referred to by most contemporary scientific musical texts. Following eight years in preparation this book was published in 1862. Three additional German editions followed, the last in 1877. Ellis made two English translations, the second in 1885. Thus, this book and the experimental explanations behind it were constructed in an age without oscilloscopes, audio oscillators, or computers. Indeed, without practical alternating current electricity. This is an important book, but from a different era.

Sources of Illustrations

Kurt Sperry drew or adapted most of the illustrations in *The Structure of Musical Sound*.

page 29 (Figure 2.2) Adapted from C.G. Conn. Ltd.

page 72 (Figure 3.1) Adapted from *The Science of Sound*, by Thomas D. Rossing, Addison-Wesley Publishing Company, 1982, page 90, and *Science and Music*, by Sir James Jean, Cambridge University Press, 1937; Dover Publications, Inc., 1968, page 227.

page 88 (Figure 5.1) Adapted from The Physics of the Piano in *The Physics of Music*, by E. Donnell Blackham, W.H. Freeman and Company, San Francisco, 1978, pages 25 and 26.

page 91 (parts of violin, Chapter 5) Adapted from *Horns, Strings And Harmony*, by Arthur H. Benade, Anchor Books, Doubleday Company, Inc., 1960, page 138, and from *The Physics of Musical Sound*, by Jess J. Josephs, Van Nostrand Reinhard and Company, 1967, page 97.

page 94 (guitar bracing, Chapter 5) Figure 10.23, page 202 from *The Science of Sound*, by Thomas D. Rossing, Copyright© 1990 Addison-Wesley Publishing Company, Inc., Reprinted by permission of Pearson Education, Inc.

Sources of Illustrations

page 95 (round membrane modes), Chapter 5 Adapted from *The Science of Sound*, by Thomas D. Rossing, Addison-Wesley Publishing Company, 1982, page 244.

page 156 (saxophone with keys removed, Chapter 9) Reproduced with permission from *Horns, Strings, and Harmony*, by Arthur H. Benade, Anchor Books, Doubleday Company, 1960, page 217. Copyright© Educational Services Incorporated.

page 164 (bottles as Helmholtz resonators and spring-bob models, Chapter 10) Adapted from *Horns, Strings, and Harmony*, by Arthur H. Benade, Anchor Books, Doubleday Company, 1960, pages 144, 145.

page 172 (Manfred Schroeder, Chapter 11) Courtesy of Professor Schroeder, copyright© Bell Laboratories.

page 189 (area of hearing, Chapter 11) Adapted from *Science and Music* by Sir James Jean, 1937, Cambridge University Press, Dover Publications, Inc. 1968, page 227.

page 192 (Figure 11.6, Chapter 11) Adapted from *The Science of Musical Sound*, by John R. Pierce, W.H. Freeman and Company, 1983, page 135.

pages 199–206 (New York Times article, Chapter 11) New York Times, March 31, 1991, by Tom Manoff. Reproduced by permission from Tom Manoff.

Index

DEMONSTRATION is abbreviated DEMO. QUESTION is abbreviated QUEST. Definitions of technical terms are found in the Glossary (pp. 317–330), which is not indexed here.

A

A_4, a_4, 29 fig. 2.2
chronology of pitch, 80 table 4.1
absorption, 229–30, 230 DEMO XXV
  table of coefficients, 232 table A.1
accordion
  free reed, 140 fig. 7.1, 141
  picture, 159
acoustic terms
  disagreement, perceptions of, 171
  music, perceptions of, 171
  music in rooms, scientific measurement of, 173
  rooms, scientific measurement of, 173
  See also specific terms;
    reverberation time; delay time; diffraction; reflection; absorption
active acoustics, 194–207
  using computers to make sound, 196, 197 fig. 11.8, 237 fig. B.1, 246 fig. B.2, 247 fig. B.3, 253 fig. B.5
  using computers to change rooms, 254–55
air
  compressed, 212
  density of, 223, 224
  elasticity of, 13 DEMO VI
  Ideal Gas Law, 215–17, 220–21
  Kinetic Theory Of Gasses, 217–20
  mechanical model of, 213 DEMO XXIV
  medium for sound waves, 3
  molecules, 213 DEMO XXIV, 215
  speed of sound in, 25, 224
air molecule
  distance between, 212, 224, 226
  simple sphere, 215
  size, 224
alternate fingering
  for trumpet, 147 fig. 8.1
analogue-to-digital conversion, ADC, 241, 243
  sampling rate, 240–41
antinode
  in standing waves
    bars, 100–01
    cavities, 115–16, 122 DEMO XVIII
    room, halls, 174 fig. 11.3, 175–78, 186 QUEST 11.4–11.5, 186, 187 QUEST 11.6, 289–90
    Slinky, 106–11 DEMO XVI
    strings, 39–40 DEMO XI, 49, 51
attack, 190, 252, 253 fig. B.5
  identifying instrument, 190
audio oscillator, 14 DEMO VII, 21–23 DEMO IX, 84–85 DEMO XIV, 122–23 DEMO XVIII
  pitch and loudness, 14, 21–23 DEMO IX

- 335 -

Index

B

Bach, J.S., 79
  even tempering, 79
  The Well-Tempered Clavier, 79
band pass filter, 246, 247 fig. B.3
banjo, 84
  modes of membrane, 95–97
  picture, 93
  sound compared to guitar, 95
bassoon
  conical cavity, 155
  picture, 155
  reed, 140 fig. 7.1
beat frequency, 61
  value, how to calculate, 61–62
  See also beats
beats
  from near-equal frequency sounds, 59–61
  loudness variations, 59, 61
  sonic, 59–62
  visual, using strobe light, 10 DEMO IV
  See also beat frequency
bell
  in brass instruments, 148, 151–53
  not needed in woodwinds, 156
  reduction of sharp musical intervals, 151–53
  reflection of sound wave at end of cavities, 137, 151–52
bell jar, 14–15 DEMO VII
  See also vacuum
Benade, Arthur H., 165, 314, 331, 332
  Fundamentals Of Musical Acoustics, 332
  Horns, Strings and Harmony, 331
block diagram
  example, 237 fig. B.1, 246 fig. B.2, 253 fig. B.5
  process design aid, 256 QUEST B.4, 310–11

boundary conditions, 115–17, 118–20
  pressure in cavities, 157
brass instruments
  bell, 151–53
  ideal conical, 145–46
  pedal note, 147
  real, 148
  slides, 146
  valves, 146
bridge
  banjo, 93, 95
  guitar, 93, 94
  pianoforte, 88 fig. 5.1, 89
  violin, 91–92

C

cavities
  conical, 136–37
  cylindrical, open at both ends, 119, 121 DEMO XVII, 122 DEMO XVIII, 138
  cylindrical, open at one end, 115, 130–31, 139 DEMO XX
  rooms, halls, 185
  soundboxes, 92, 93, 94–95
cavity-controlled oscillator, 126
  See also brass instruments; reeds; woodwinds
computer-controlled audio electronics, C-CAE, 195, 236
  changes to room acoustics, 254–55
  changes to sound, 196–97, 197 fig. 11.8, 237 fig. B.1, 246 fig. B.2, 253 fig. B.5
changes, 104
Chladni, Ernst F.F., 85
Chladni figures, 85, 95
  round membranes, 95
  See also antinodes; nodes; resonance; standing waves

Index

chromatic scale, 33
  names of tones in, 34
  semitones in, 33, 34
  See also even tempering

clamping, 48
  at atmospheric pressure, 115, 158
  to cause nodes, 48–49, 157
  to prevent modes, 48–49, 49
  QUEST 3.2, 101
  by register key in woodwinds, 158
  by tone holes in woodwinds, 119
  by supports of vibrating bars, 101
  See also nodes

clarinet
  picture, 155
  playing registers, 284
  reed, 140 fig. 7.1, 222
  register key, function, 119
  register key, location, 119, 158
  QUEST 9.1, 281–86, 282, 284, 285
  tone holes, 119, 283
  See also cylinders closed at one end

combination frequency, 62
  adding musical intervals, 63, 70
  DEMO XII
  changing pitch, 67
  changing timbre, 65, 67
  ear-brain system, 62
  non-linear devices, 62

complex vibration, 49
  strings, 49–51
  See also superposition

complex wave, 51–58
  new frequency present, 58
  shape and phase, 52–55, 56
  See also superposition

computer
  clock rate, 238
  hard-wired operations, 245
  parallel computing, 243, 245, 248
  programming, 238
    See also block diagram; Fourier transform; Fast Fourier Transform

  speed, 238, 257
  use in computer-controlled audio electronics, 195–97, 197 fig. 11.8, 237 fig. B.1, 246 fig. B.2, 247, 253 fig. B.5

concertina, 159–60
  picture, 159
  See also free reeds

cones, 136
  modes, 137
  See also bassoon; oboe; recorder; saxophone

continuity, 128

Cooley, J.W., 239, 315
  See also Fast Fourier Transform

cycle, 6, 18, 19, 41, 42–43
  in beats, 60–61
  in complex waves, 53–58, 59
  QUEST 3.4, 64, 66, 269–70
  definition, 20
  length of, 24, 41, 132, 138
  not in noise, 18
  in standing waves, 41, 107, 109
  in traveling waves, 24

cylinders closed at one end, 115
  frequencies of modes, 118, 133–35
  modes, 116–18
  wavelengths of modes, 118, 132
  See also clarinet

cylinders open at both ends, 136, 138
  frequencies of modes, 121 DEMO XVII, 138
  modes, 138
  wavelengths of modes, 138
  See also flute

D

decibel, dB, 73
  formula for calculating, 73
  and phon, 74 QUEST 3.8, 271
  See also loudness

Index

delay time, 173
  definition, 192
  and echoes, 193
  and perception of musical sound, 192–94, 193 fig. 11.7
  property of room, 194
  See also direct sound; indirect sound

density
  of air, 223

diffraction, 173, 211, 228
  in halls and rooms, 229
  sound waves, 229, 299
  water waves, 228, 228 fig. A.1

digital-to-analogue conversion (DAC), 242

direct sound, 192, 192 fig. 11.6
  from loud speakers, 249, 254, 255 QUEST B.2–B.3, 302–10, 311
  from musicians, 193, 193 fig. 11.7, 253, 255 QUEST B.2–B.3, 302–10

Dolmetsch, Arnold, 165–66, 166 QUEST 10.2

Doric mode, 35

dynamic range
  loudness, 250

E

ear-brain system, 62, 72
  create combination frequencies, 62
  non-linear amplifier, 62
  range of hearing, 72, 72 fig. 3.1, 73, 189
  threshold of hearing, 72, 72 fig. 3.1, 73, 189
  threshold of pain, 72, 72 fig. 3.1, 73, 189

echo
  and room acoustics, 193, 193 fig. 11.7
  See also delay time

ensemble, 173

envelope, 108, 116, 275

even tempering, 29 fig. 2.2, 77, 78, 79
  ability to play in all keys, 75, 77
  compared to just tempering, 79
  required by pianoforte, 77
  See also Bach, J.S.

F

Fast Fourier Transform (FFT), 239, 248–49
  speed of, 239
  See also Cooley, J.W.; Tuckey, J.W.; Fourier Transform

flute, 136
  air reed, 140 fig. 7.1, 141
  modes, 138
  picture, 155
  See also cylinders open at both ends

formulas
  frequencies of standing waves in conical cavities, 137–38
  frequencies of standing waves in cylinders open at both ends, 121, 138
  frequencies of standing waves in cylinders open at one end, 118
  frequencies of standing waves in strings, 45–46
  wavelengths of standing waves in cylinders open at both ends, 120
  wavelengths of standing waves in cylinders open at one end, 118
  wavelengths of standing waves in strings, 45

Fourier analysis, 239

Fourier coefficients, 239

Fourier, Jean Baptiste, 239

Fourier transform, 239
  See also Fast Fourier Transform

free reed, 141–142, 159–61
  See also accordion; concertina; harmonica

Index

French horn. *See* horn

frequency, 21, 25, 31, 35 table 2.1, 37 QUEST 2.9–2.11, 80 table 4.1, 265–66
  modes of round membranes, 96
  modes of thick strings, 101
  and periods, 21, 24
  range, voices and musical instruments, 29 fig. 2.2
  resonant, 114
  units, 24–25
  See also period; pitch

Fundamentals Of Musical Acoustics, 332
  See also Benade, Arthur H.

G

glockenspiel, 100

guitar
  bracing, 94
  Helmholtz resonator, 94
  picture, 93
  sound box, 94

H

hard-wired operations, 245, 247

harmonica, 159–61
  picture, 159

harmonics, 40, 46, 59
  integer multiples, 40
  names of modes, 40
  See specific musical instruments
  See also mode

hearing
  range of frequencies, 72 fig. 3.1, 189

Helmholtz, Herman L.F.
  On The Sensations Of Tone As A Physiological Basis For The Theory Of Music, 314, 332
  See Helmholtz resonator

Helmholtz resonator, 162, 162–63 DEMO XXII, 163 DEMO XXIII
  ocarina, example of, 161–62
  spring-bob model, 163–65

hertz, Hz, 24–25
  frequency unit, 24–25

Hertz, Heinrich Rudolph, 24

high, 4, 7–9, 211
  concentration of air molecules, 8–9
  creation, 128, 131
  pressure, 8–9, 19
  relative to atmospheric pressure, 19, 20, 211
  See also pressure; puff

horn
  percentage of cylindrical and conical tubing, 148
  picture, 148

Horns, Strings and Harmony, 331
  See also Benade, Arthur. H.

Hutchins, Carleen Maley, 314, 331
  The Physics of Music, 331

I

Ideal Gas Law, 215–17
  equation, 216
  See also mechanical model of a gas; pressure; temperature

impedance
  in bells of brass instruments, 233
  QUEST A.12, 300
  definition, 233

indirect sound, 192–93, 302–03, 303–10
  definition, 192 fig. 11.6
  from loudspeakers, 311
  from reflections, 192 fig. 11.6
  time of arrival, 193

interference. *See* superposition

inverse-square law, 227, 298–99

Index

J

just tempering, 35 table 2.1, 36, 75, 79
  musical intervals, 75
  pianos abandon, 75–77
  See also musical intervals

K

kettle drum
  round Chladni modes, 95, 99
  picture, 99
  tension of membrane, 99
key
  of musical piece, 34
  register, in clarinet, 158 QUEST 9.1, 281–86
  register, in woodwinds, 157–58
Kinetic Theory of Gases, 217–20
  absolute zero temperature, 219–20
  equation of a root-mean square speed, 219, 223
  See also temperature

L

lasso d'amore, 121 DEMO XVII
lip-reed, 140–41, 140 fig. 7.1
  player controls, 141, 142
logarithmic scale, 72–73, 72 fig. 3.1, 89
  See also decibel, dB
longitudinal wave, 105–11
  standing
    in air, 106
    Slinky, 106–11 DEMO XVI
  traveling
    Slinky, 11 DEMO V
  in wind instruments, 130–31, 136–37
loudness
  hearing, 21–23 DEMO IX, 71–72, 72 fig. 3.1
  and pressure, 71–72, 72 fig. 3.1
  and reverberation time, 250–55, 251 fig. B.4, 252 fig. B.4.A, 253 fig. B.5, 255 QUEST B.2–B.3, 302–10
  See phon,
loud speaker
  use of, 14 DEMO VII, 21–22 DEMO IX, 70 DEMO XII, 84 DEMO XIV, 122–23 DEMO XVIII
  See also transducer
low, 4, 7–9, 211
  concentration of air molecules, 8–9
  creation of low, 127, 130–31, 141
  partial vacuum, 7–8, 211
  pressure, 8–9, 19
  relative to atmospheric pressure, 19, 20, 211
  See also pressure

M

macroscopic property
  pressure, 214
  temperature, 219
  vacuum, 212
Major mode, 33–34, 34–35 QUEST 2.5–2.6, 36, 264
  definition, 34 chart 2.1
  See also semitone
Manoff, Tom
  New York Times, acoustics article, 198–206
marimba, 100
mathematics, 103–04
  See changes
mechanical model of a gas, 213 DEMO XXIV
  hard spheres, 214
  See also pressure; temperature
microphone
  operation of, 7 DEMO III

Index

use of, 5–6 DEMO II, 17 DEMO VIII, 21–22 DEMO IX, 122 DEMO XVIII
See also transducer

microscopic property, 214, 219

Minor mode, 33–34, 35 QUEST 2.7, 36, 265
definition, 34 chart 2.1
See also semitone

mode
Chladni figures, 85
circular membranes, 95–96
cones
frequencies, 137
cylinders closed at one end, 116–18
frequencies, 118
cylinders open at both ends, 119–20
frequencies, 121
fundamental, 40, 41
harmonics, 40, 41, 46
musical definition, 33
and musical intervals, 47 table 3.2
real brass instruments, 150–51
scientific definition, 33
thick strings, 101
vibrating string, 40
frequencies, 41, 45, 46
wavelengths, 44 table 3.1, 45
See also Major mode; Minor mode

model
making, 3
mechanical, of a gas, 213–14 DEMO XXIV, 215
proposed for sound, 4, 16 (incorrect)

musical instruments
vs. noise makers, 17 DEMO VIII
See under specific kinds

musical intervals
created by combination frequencies, 63–71
created by superposition, 63–71

frequency ratios for even tempering, 77–78
frequency ratios for just tempering, 35 table 2.1
in modes, 47 table 3.2
names, 35 table 2.1, 47 table 3.2

musical sound
description of, 16, 18, 19
not noise, 17–19 DEMO VIII
sampling rate
analogue-to-digital conversion, 240
Fast Fourier Transform, 243

N

near wall effects, 175

New York Times
acoustics, 198–206

node
in Chladni figures, 85, 95
in rooms, 174–75, 177–78, 186
in standing waves, 122–23 DEMO XVIII
caused by oppositely traveling waves, 113, 113 QUEST 7.2, 274–76
caused by register key, 157–58, 158 QUEST 9.1, 281–85
in strings, 40
in thick strings, 100–01

noise
definition, 18–19
not musical sound, 17–19 DEMO VIII, 18

non-linear
definition, 62
See also combination frequencies; hearing

Index

O

oboe
 conical, 155
 reed, 140 fig. 7.1
 picture, 155

ocarina, 161–62
 picture, 161
 See also Helmholtz resonator

octave
 even-tempered, 78
 just-tempered, 35 table 2.1
 lettered notes, 32 fig. 2.3
 names of tones, 33, 34
 pianoforte, 32 fig. 2.3

oscilloscope
 description, 5–6 DEMO II
 use of, 5–6 DEMO II, 17–19 DEMO VIII, 21–22 DEMO IX, 122–23 DEMO XVIII

P

pan pipes, 140 fig. 7.1

parallel computing, 243, 245, 248

Pascal
 pressure unit, 73, 220, 221

Pascal, Blaise, 220

pedal note, 147

percussion instruments
 kettle drum
  round Chladni figures, 99
 thick strings
  modes, 101
  supported at nodes, 101
 See also vibraphone

period, 21
 See also frequency

phase
 completely out of phase, 54
 effect on timbre, 56
 in phase, 52, 57 QUEST 3.3, 186, 266–68
 out of phase, 56, 57 QUEST 3.3, 186, 266–68
 shape of complex wave, 52–56
 tone change due to phase change, 56

phon, 72 fig. 3.1, 74 QUEST 3.8, 271
 See also loudness

physical object, 259, 260, 261
 integer behind symbol, 261
 no italics, 260

physical quantity, 259, 260
 italics, 259
 number before symbol, 261
 subscripts and superscripts, 261
 units, 259

piano, pianoforte, 86
 abandons just tempering, 75
 naming tones of, 32–33
 parts of, 88 fig. 5.1, 89
 range, 29 fig. 2.2
 short length of low pitch strings, 89 QUEST 5.2, 272
 tempering pitches of, 77, 79

Pierce, John R.
 The Science of Musical Sound, 314, 331

pitch, 21
 and frequency, 21 DEMO IX, 71
 ghostly, 70 DEMO XII
 name of piano key, 31, 32 fig. 2.3
 notation using staff, 31
 range for musical instruments, 29 fig. 2.2
 semitone interval, 33
 subjective, 71

pressure, 220
 absolute, 221
 atmospheric, 220
 concentration of gas molecules, 8, 215, 223
 gauge, 221

- 342 -

Index

high and low, 222–23
highs and lows in musical sound, 8, 19, 20, 222
and loudness, 72 fig. 3.1
macroscopic property, 212, 214
speed of gas molecules, 223
table, 221
units, 73, 216, 220
See also Ideal Gas Law; mechanical model of a gas

puff
in brass, when added, 124, 142, 143
real interval of puff, 143–44
sustaining sound wave, 123–24
woodwinds, when added, 124, 143
See also cavity-controlled oscillator; high

R

range of musical instruments
pitch and frequency, 29 fig. 2.2
recorder
air reed, 140 fig. 7.1
conical cavity, 155
picture, 155
reeds, 140 fig. 7.1
air, 141, 142
cane, 141, 222
free, 141, 159–60
lip, 141, 142
real, 142–44
See also cavity-controlled oscillator; puff
reflection
caused by impedance changes
at bell, 151–53, 153 QUEST 8.6, 280–81
in rooms. *See* delay time; resonance; reverberation time
at tone holes, 156
in strings, 112
causing resonances, 173–75

from closed ends, 112, 151–52
from open ends, 123, 127, 130, 137, 138, 151–52
of highs, 127, 151–52, 186
loss of sound, 227–28
of lows, 186
and reverberation time, 173–75
water waves, 27–28, 27 fig. 2.1

register key
definition, 157–58
location of, 157–58, 158 QUEST 9.1, 281–86

resonant frequency. *See* frequency

reverberation time, RT, 188–91
changing
computer-controlled audio electronics (C-CAE), 190, 246 fig. B.2, 253 fig. B.5
room changes, 254–55, 255 QUEST B.2–B.3, 302–10
definition
hearing dependent, 188, 190
scientific, 173, 188
and loudness, 250–55, 251 fig. B.4, 252 fig. B.4.A
recipes for changing, 191
and reflectors, 253–55

ripple tank, 25–28, 25–28 DEMO X
to demonstrate properties of traveling waves, 25–28 DEMO X
to show diffraction, 228–29, 228 fig. A.1

root mean square speed, 217, 219, 223
and average speed, 223

Rossing, Thomas D.
explanation of piano-forte action, 89
Musical Acoustics, Selected Reprints, 332
The Science of Sound, 332

Index

S

St. Martin's Hall, 170 fig. 11.2
Sax, Adolphe
  saxophone, 154
  tenor trombone, 154 fig. 8.2
saxophone
  picture, 155
  stripped of keys, 156
scales
  chromatic (12 tone), 33, 34, 78
  comparison of just and even tempered, 79
  even-tempered, 77, 78
  just-tempered, 75 (failure)
  Major mode, 34 chart 2.1, 78
  Minor mode, 34 chart 2.1
  names of tones, 33, 34
  scale (8 tone), 33, 34 chart 2.1
Schroeder, Manfred, 172, 315
  acoustic properties of rooms, 172–73
  Journal of the Audio-Engineering Society, 315
science
  model making, 3
  theory and experiment, 16
  See specific scientific terms
semitone, 33
  chromatic scale, 33, 34
  even tempering, 77
  intervals in Major and Minor scales, 34 chart 2.1
  names, 33, 34
  sharps and flats, 33
side hole. *See* tone hole
slides
  in brass, 145, 151
  trombone, 148, 151
Slinky, 11 DEMO V, 106–07 DEMO XVI
  model of longitudinal wave, 105–06
  model of traveling wave, 12
  standing wave in, 106–07 DEMO XVI

Smith, Cyril Stanley, 166
soundboard, 84, 84 DEMO XIV, 85, 86
  banjo, 94, 95
  good shape for, 86
  guitar, 94
  need for, 83, 83 DEMO XIII
  pianoforte, 88 fig. 5.1, 89
  See Chladni figures
soundbox, 83, 86
  good shape for, 86
  guitar, 94
  Helmholtz resonators, 86
  need for, 83
  violin, 92, 94
sound waves, 9, 11, 24, 226
speaker hole. *See* register key
speed of wave
  air, 25, 224, 235
  electrical signals, 235
  strings, 45
  water, 25
spring-bob oscillator
  mechanical, 125–26 DEMO XIX
  model for Helmholtz resonator, 163–64, 165
  sustaining oscillations, 125–26 DEMO XIX
  See cavity-controlled oscillator
staff
  for writing music, 29 fig. 2.2, 31, 103, 104, 104 QUEST 6.1, 147 fig. 8.1, 274
standing waves, 39, 179 fig. 11.5
  in air in cavities, 122–23 DEMO XVIII
  boundary conditions, 115–16
  conditions for, 106, 106–07 DEMO XVI
  longitudinal, 106, 106–07 DEMO XVI
  modes, 39–40 DEMO XI
  resonant frequency, 114
  in Slinky, 106–07 DEMO XVI

Index

in strings, 39–40 DEMO XI
in thick strings, 100–01
in rooms, 174 fig. 11.3, 175 (false), 178 fig. 11.4, 179 fig. 11.5, 183–84

strings
banjo, 93
clamping, 48, 101
guitar, 93
name of type of musical instrument, 84
need for additional loudness, 83–84 DEMO XIII
pianoforte, 86–88
thin and thick, 39, 86 (thin), 100 (thick)
vibrating, 39
violin, 92

strobe light
visual beats, 10
use of, 10 DEMO IV

subscripts
use of, 261

superposition
to create beats, 59–61
to create complex waves, 52, 53–59 (examples)
to create standing waves
in strings, 50
in water, 25 DEMO X
definition, 28

superscripts
use of, 261

sweet potato. *See* ocarina

T

Teatro alla Scala, 169, 170 fig. 11.1
temperature
absolute zero, 216, 219
Celcius scale, 220
Kelvin scale, 220
macroscopic property, 219
table of typical, 220

See also Ideal Gas Law; Kinetic Theory of Gases

tempering
even, 29 fig. 2.2, 36, 77, 79
pianoforte and need for, 77
just, 35 table 2.1, 36, 75, 79

Thomson, William, Lord Kelvin, 220

threshold
audibility
and background noise, 250, 251, 252, 254
frequency dependent, 72 fig. 3.1, 189–90
hearing, 72 fig. 3.1, 189–90, 250, 251, 252, 254
pain, , 72 fig. 3.1

timbre, 74
affected by reed, 144
attack, 190
mode content, 71, 74
and phase, 56, 313
and reverberation time, 190, 250

timpani. *See* kettle drum

tone hole
and reflection, 156
shortens wind instrument, 119
size of in saxophone, 156

transducer
definition, 6
electric signal-to-soundwave
loudspeaker, 235, 246 fig. B.2, 247 fig B.3, 311
soundwave-to-electric signal
microphone, 6, 235, 247 fig. B.3

transverse wave
definition, 105
in strings, 105
in water, 25–28 DEMO X

traveling waves
in Slinky, 11–12 DEMO V
in strings, 11–12 DEMO V
in water, 25–28 DEMO X
sound, model of, 7–9

Index

trombone
  percentage of cylindrical and conical tubing, 148
  picture, 148

trumpet
  ideal conical, 146
  percentage of cylindrical and conical tubing, 148
  picture, 148
  range of tones
    ideal conical, 147 fig. 8.1
    real, 151

Tukey, J.W., 239
See also Fast Fourier Transform

tuning fork
  increasing loudness with sound board, 83 DEMO XIII
  operation
    to make sound, 4 DEMO I, 5 DEMO II, 17 DEMO VIII, 83 DEMO XIII
    shown with strobe light, 10 DEMO IV

V

$(v^2)_{ave}$. *See* root mean square speed

vacuum
  absence of sound in, 14–15 DEMO VII
  definition, 211
  macroscopic property, 212
  partial, 211, 221
  table of pressures, 221

valves
  alternative fingerings, 147, 147 fig. 8.1
  in brass to lengthen tubing, 146

vibraphone, 100
  picture, 100

violin, 91–93
  bass bar, 91, 92
  bridge, 91, 92
  loudness, 92, 93
  soundbox, 92
  sound post, 91, 92, 93

violin family
  picture, 90–91

W

wave length, 23–25
  and boundary conditions for wind instruments, 116–21
  length of cycle, 24, 109
  standing wave, 39–40 DEMO XI, 44, 109, 115, 116–21
  traveling wave, 24, 43
  wave length-frequency-speed relationship, 25

The Well-Tempered Clavier, 79
  test of even tempering, 79

woodwinds
  ideal and real, 155
  picture, 155
  See also specific instruments

X

xylophone, 100